HISTORY OF
HYDROLOGY

History
of hydrology

Asit K. Biswas

Research Coordination Directorate
Department of Environment, Ottawa, Canada

1972

NORTH-HOLLAND PUBLISHING COMPANY
AMSTERDAM – LONDON

Library of Congress Catalog Card Number: 69–18384
ISBN North-Holland: 7204 8018 3
ISBN American Elsevier: 0444 10025 3

PUBLISHERS:

NORTH-HOLLAND PUBLISHING COMPANY – AMSTERDAM
NORTH-HOLLAND PUBLISHING COMPANY, LTD. – LONDON

SOLE DISTRIBUTORS FOR THE U.S.A. AND CANADA:

AMERICAN ELSEVIER PUBLISHING COMPANY, INC.
52 VANDERBILT AVENUE
NEW YORK, N.Y. 10017

First edition 1970
Second printing 1972

PRINTED IN THE NETHERLANDS

Preface

According to Comte, "no one can be really master of any science unless he studies its special history", but, unfortunately, the history of the development of the science of hydrology has been a most neglected subject. Even in the closely related field of hydraulics, only one serious study is available: *History of Hydraulics* by Hunter Rouse and Simon Ince. Thus, it is not surprising that hydrologists have very little knowledge of their heritage, and historically erroneous statements can frequently be seen in hydrologic literatures. Early 1964, I started to read the works of early hydrologists primarily to satisfy my own curiosity. As time progressed, the subject became more and more fascinating, and I wrote a few papers on certain specific aspects of history of hydrology – notably in the journals of the Royal Society of London, American Society of Civil Engineers, International Association of Scientific Hydrology and the Society for the History of Technology. The papers, somewhat to my surprise, were enthusiastically received by both hydrologists and historians of science and technology. This enthusiastic response, coupled with the requests from fellow hydrologists from all the world, prompted me to put my work in a book form.

One of my major difficulties was a universally accepted definition of hydrology. Unfortunately, no such definition exists, and even if it did, considering the present-day tendency of multi-disciplinary approach to subjects, any such definition is bound to be overinclusive. For example, where does hydrology end, and two closely associated endeavours, hydraulics and meteorology, begin? In the present case, discharge measurement techniques, both equipments and formulae, have been considered as hydrologic events whereas other aspects of open channel flow are assumed to be within the domain of hydraulics. It can be argued, with some justification, that sedimentation, weather modification, soil physics and some aspects of oceanography, may be included within the field of hydrology, but to keep the book within manageable proportion, I have excluded them, and I do not propose to present a lengthy argument for or against such a step.

During the course of my research, I visited a number of museums and university libraries in Europe and North America. In the text, wherever possible, I have acknowledged their specific assistance. Many distinguished hydrologists and historians of science and technology, from all over the world, have freely given their advice and aid. It is impossible for me to express my grateful appreciation to them. But special mention should be made of continued interest and assistance from Professor Ray K. Linsley, Stanford University; Professor Ven T. Chow and Professor George White, University of Illinois; Professor J. C. I. Dooge, University College, Cork, Ireland; Professor A. Volker, Rijkswaterstaat, Den Haag, The Netherlands; Sir Harold Hartley, F.R.S., Central Electricity Generating Board, London; Professor Aurèle La Rocque, Ohio State University; Professor Gunther Garbrecht, Middle East Technical University, Ankara, Turkey; and late Professor William Fraser and Professor D. I. H. Barr, University of Strathclyde, Glasgow. My greatest debt, however, is to Arthur H. Frazier, former Chief, Division of Field Equipment, U.S. Geological Survey, without whose co-operation this book would never have been written. Finally I would like to express my appreciation to the staff of the North-Holland Publishing Company, especially Mr. A. T. G. van der Leij, for their unfailing cooperation.

Ottawa, Canada,   Asit K. Biswas
March 6, 1970.

Acknowledgements

The author is deeply grateful to the following publishers for granting permission to quote from the designated books published by them.

G. Bell & Sons, London; Aristophanes, The clouds, translated by B. B. Rogers, 1952.

Basil Blackwell, Oxford; K. Freeman, Ancilla to pre-socratic Philosophers, 1948.

Clarendon Press, Oxford; Aristotle, Meteorologica, translated by E. W. Webster, 1952; Plato, Timaeus, translated by B. Jowett, 1953.

Hafner Publishing Co., New York; P. Perrault, On the origin of springs, translated by A. LaRocque, 1967.

Harvard University Press, Cambridge, Mass.; M. R. Cohen & I. E. Drabkin, Editors, A source book in Greek science, 1948.

Macmillan & Co., London; L. A. Seneca, Physical science in the time of Nero, translated by J. Clarke, 1910.

Miss G. M. A. Richter and Oxford University Press, London; Leonardo da Vinci, The literary works of Leonardo da Vinci, edited and translated by J. P. Richter, 1939.

Russell & Russell, New York; Lucretius, Roman poet of science: Lucretius, De Rerum Natura, translated by A. D. Winspear, 1955.

University of Illinois Press, Urbana; B. Palissy, The admirable discourses of Bernard Palissy, translated by A. LaRocque, 1957.

Contents

1

Hydrology prior to 600 B.C.

INTRODUCTION

When and where did the science of hydrology begin? It is difficult to answer, as the roots of modern hydrology lie deeply buried in antiquity. From the beginning, man realized that water is essential for survival, and hence it is not surprising that evidences of the earliest civilizations have been found along the banks of rivers: the Tigris and Euphrates in Mesopotamia, the Nile in Egypt, the Indus in India, and the Huang-Ho (Yellow River) in China. Gradually, they developed their water supply systems, constructed dams and levees, made channel improvements, and dug canals for drainage and irrigation. The presence of these structures proves that man had some knowledge of water, its powers and limitations – although it was admittedly not very scientific. The first hydrologic principles were extremely crude, but in the beginning man was primarily interested in controlling nature; and only later, during the Hellenic Civilization[1] (around 600 B.C.), did he try to understand nature. It may be said that one cannot treat a branch of science as such until a certain degree of development has taken place, but who will define and measure that degree? To paraphrase the great historian of science, George Sarton, a 2-in. high *Sequoia gigantea* may not be very conspicuous, but it is still a *Sequoia*. After all, when the first primitive mathematician realized that there was something similar about three palm trees and three donkeys, how abstract was his thought?

ANCIENT 'HYDROLOGIC' WORKS

The three major civilizations that flourished some 4000 years ago were those of the Egyptians in the Nile Valley, the Sumerians in Mesopotamia, and the Harappans in the Indus Valley.[2] Much is known about the Egyptian and the Sumerian civilizations, but little about the Harappans; yet theirs embraced an area more extensive than either of the others. In contrast to the abundance of written records available of the Egyptians and the Sumerians, the history of the Harappans must be gleaned from the archaeological findings at Harappa and Mohenjo-daro, mainly because written records are scarce and are yet to be deciphered (the civilization was excavated only in the 1920's). It has been argued recently that the Harappan civilization came to an end in a catastrophic flood of the River Indus due to tectonic disturbances.[3]

The Chinese civilization grew up on the banks of the Huang-Ho river, but their contribution to hydrology, especially prior to 600 B.C., is not as significant[4] as that of the other three. All four civilizations had, in general, similar geographical conditions: they were based on major rivers, rainfalls were scanty, summer temperatures were high, and the phenomena of river fluctuations were comparable.

A brief review of the development of early hydrologic and hydraulic engineering around the world is presented in this chapter, and a chronology of hydrologic engineering is shown in the table.

'KING SCORPION' AND KING MENES

Probably one of the earliest evidences of hydrologic work to have been accomplished can be found in the drawing of an imperial macehead held by the protodynastic 'King Scorpion',[5] a ruler who acquired his name from the scorpion appearing in front of him (fig. 1). In the drawing he wears the white crown of Upper Egypt and is probably 'cutting the first sod' of an irrigation ditch, a ceremony inaugurating the inundation which has continued to be practiced into the nineteenth century[6] at the festival of the 'day of breaking the river'. In front of the king is a man holding a basket (probably filled with seeds), and beyond him is another man who is holding

TABLE

A chronology of recorded hydrologic engineering prior to 600 B.C.

Date* (B.C.)	Event
3200	Reign of King Scorpion; first recorded evidence of water resources work.
3000	King Menes dammed the Nile and diverted its course.
3000	Nilometers were used to record the fluctuations of the Nile.
2850	Failure of the Sadd el-Kafara dam.
2750	Origin of the Indus Valley water supply and drainage systems.
2200	Various waterworks of 'The Great Yü' in China.
2200	Water from spring was conveyed to the Palace of Cnossos (Crete). Dams at Mahkai and Lakorian in Persia.
1950	Connection of the Nile River and the Red Sea by a navigational canal during the reign of Seostris I.
1900	Sinnōr constructed at Gezer (Palestine).
1850	Lake Moeris(?) and other works of Pharaoh Amenemhet III.
1800	Nilometers at Second Cataract in Semna.
1750	Water codes of King Hammurabi.
1700	Joseph's Well near Cairo, nearly 325 ft in depth.
1500	Two springs joined by a sinnōr in the city of Tell Ta'annek in Palestine.
?	Marduk Dam on the Tigris near Samarra, destroyed in 1256 A.D.
1300	Irrigation and drainage systems of Nippur. Quatinah Dam on the Orontes River in Syria constructed under the reign of Sethi I or Ramses II.
1050	Water meters used at Oasis Gadames in North Africa.
750	Marib and other dams on River Wadi Dhana in Yemen.
714	Destruction of qanāt systems of Ulhu (Armenia) by King Saragon II. Qanāt system gradually spread to Persia, Egypt, and India.
690	Construction of Sennacherib's Channel.
600	Dams in the Murghab River in Persia, destroyed in 1258 A.D.

* In the absence of accurate information, many of these dates are approximate.

ears of corn. Below him are workmen who are adding the final touches to the canal. The king who has a hoe in his hand is about three times larger than the other men around him, which probably is an indication of his authority and power. The Pharaoh ruled sometime during the fourth millennium, probably around 3200 B.C. Little additional information is known about this mysterious king with his characteristic scorpion and his seven-petalled lotus insignia.

Figure 1. 'King Scorpion' cutting the first sod of an irrigation canal (by courtesy of Ashmolean Museum, Oxford).

King Menes ruled around 3000 B.C. He built his capital at Memphis, and was legendarily the first of the Pharaohs. According to the historian Herodotus[7] he dammed the Nile about $12^1/_2$ miles south of Memphis at Kosheish, and diverted the course of the river to a newly dug channel between two hills. The gravity dam seems to have had a maximum height of about 50 ft and a crest length of some 1470 ft.[8] The new capital of Memphis was then built on the old fertile river bed. Later, he excavated a lake to the north and west, of the new town, and dug a canal to connect the lake with the River Nile. The system of water courses, *viz.*, the lake, the canal, and the river, served as a moat to protect King Menes from his enemies. Consequently, the new dam had to be carefully guarded and maintained, because had there been a breach, the entire city of Memphis would have been flooded. When Herodotus visited Egypt some 2500 years later, the dam was still guarded with the greatest care by the Persians. The Egyptian priests showed the historian a list of some 330 monarchs who followed King Menes, but they were 'personages of no note or distinction', except the last who was named Moeris (better known as Amenemhet III), of whom further mention will be made later.

SADD EL-KAFARA DAM

In 1855, Schweinfurth[9] discovered the remains of what is sometimes called the oldest dam of the world, in the Wadi el-Garawi about 18 miles south of Cairo, Egypt. The abutments of the Sadd el-Kafara ('Dam of the Pagans') are still in existence, and it is now generally agreed that the dam (figures 2 and 3) was built during the third or the fourth dynasty, sometime between 2950 B.C. and 2750 B.C. Murray[10] made detailed measurements in 1935, and

Figure 2. Sadd el-Kafara Dam (2800 B.C.) looking downstream (by courtesy of Mrs. G. Murray).

Figure 3. Facing the Sadd el-Kafara Dam (by courtesy of Mrs. G. Murray).

found that the structure was 348 ft long at the top and about 265 ft long at the base, with its crest some 37 ft from the lowest bed level. It was made up of two separate rubble masonry dams, each 78 ft thick at the base, and they were separated by a space of 120 ft along the stream bed, which was later filled in with over 60,000 tons of shingle from the river bed and the adjacent hill. The upstream and the downstream dams contained about 30,000 cu. yd. of dry rubble masonry. The upstream portion of the dam had a carefully placed facing of roughly dressed limestone blocks weighing about 50 lb. each, in steps 11 in. high. Probably the engineer was pressed for time, as the downstream portion must have been hurriedly built; this would account for the shoddy workmanship of the downstream side compared to the upstream construction.

Two notable features of the dam were that there was no provision for a spillway (was the dam supposed to be a temporary structure?), and that no mortar was used in its construction. Schweinfurth believed that the dam had been constructed to provide drinking water for the workmen and animals of an alabaster quarry located about two miles to the east, and hence its failure due to overtopping probably did not have catastrophic consequences. The capacity of the reservoir was only 460 acre ft, and the drainage area was 72 sq. miles. Assuming that the climate of 5000 years ago was similar to that of today (in fact Murray[11] has shown it to be so), the first heavy rains would probably have been enough to fill the reservoir and overtop the dam.[12]

Complete absence of sediments upstream of the dam probably indicates the dam had failed during its first flood season. Murray stated that:

'it is hard not to feel sympathy with the unknown engineer who so boldy attempted the impossible – for that.age. His was as notable a failure as that of Brunel with his 'Great Eastern' or that of Winstanley with the first Eddystone Lighthouse; but with a difference. The modern attempts were not absolute failures. They pointed the way for the successful designers of the next generation. But the Sadd el-Kafara is no landmark in the history of engineering. It was an outstanding solitary adventure which merely taught the ancient Egyptians never to attempt anything of the sort again. Its designer thought far ahead of this time. Had he made use of mortar, had he provided a spillway, had he chosen a wadi with a gentler slope, how different might have been the history of Egyptian irrigation!'[10]

The Egyptians did however build another dam during the reign of Sethi I (1319–1304 B.C.) on the Nahr el Asi (Orontes) near Homs in Syria. It was a 20 ft high and about 6560 ft long rockfill dam which is still in use.[13]

YÜ, THE GREAT

According to Matschoss,[14] the legendary hero-emperor in China Yü, the Great, was asked in about 2280 B.C. by Emperor Yau to build great waterworks, dams and dikes. He studied the rivers, and showed genius in controlling them. He was instrumental in reclaiming much land and was said to have 'mastered the waters'. So impressive was his work that after the death of Emperor Shun,

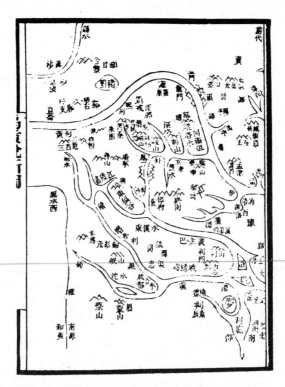

Figure 4. Diagrammatic chart of the river systems of West China from Fu Yin's *Yü kung shuo tuan* (by courtesy of Joseph Needham).

he became the new Emperor of China. In later times he became the patron of all hydraulic, irrigation, and water resources engineers. Even in the early twentieth century prayers were offered for this great engineer-emperor in all of the river temples of China. Emperor Yü, as the eminent sinologist Joseph Needham has pointed out,[15] is a legendary character, and no one knows exactly when he lived – or indeed even whether he was an actual historical person at all. Nor can one point to any specific hydraulic or hydrologic undertaking that can be attributed to the times of Yü. But anything having to do with hydrologic engineering, in medieval China, was always considered to be under the aegis of this great legendary imperial engineer. For example, figure 4 shows a diagrammatic chart of the river systems of West China,[15] from a printed work of 1160 A.D. which carries the name of Yü because of that legend.

It may be pointed out that the Chinese classified their emperors as being 'good dynasty' or 'bad dynasty' depending on whether they maintained their waterworks carefully or allowed them to fall into disrepair.[16]

PHARAOH AMENEMHET III

It is sometimes reported,[17] on the authority of Herodotus, that during the Middle Kingdom (2160–1788 B.C.) artificial lakes were used to store and control the high flood flows of the Nile. The most famous Pharaoh of this period was Amenemhet III (described as King Moeris by Herodotus) who ruled for nearly 50 years (1850–1800 B.C.).

Herodotus was much impressed by the 'artificial' Lake Moeris (now called Birket Qārūn), and according to his description it had a circumference of 450 miles, which is equal to the entire Egyptian coast line. Of elongated shape, the lake had an area of 656 sq. miles, a capacity of 40,000,000 acre ft,[18] and its maximum depth was 50 fathoms. There were two pyramids rising 300 ft above the water level at the middle of the lake, and Herodotus estimated the height of the structures to be about 600 ft.[19] During floods on the Nile, water was diverted to Lake Moeris by means of a canal. When the inundation came to an end, the stored water of the lake was returned to the Nile, and thus the storage capacity of the lake was made available for the next flooding season. Two earthen dams controlled the flow,[20] and they were cut only in times of emergency. If the dams were cut during a year of normal or low flow, the level of the Nile dropped substantially in Lower Egypt, thus making famine inevitable. These breaches were later repaired at great expense, and the labor force necessary for such work was considered excessive, even by the pyramid builders.

Herodotus also mentioned that he had heard a rumour that there was a subterranean passage from Lake Moeris to the Libyan Syrtis. This is confirmed by the Greek historian Diodorus Siculus, who visited Egypt during the first century B.C. He was much impressed by the water resources development works of the Egyptians, and according to him, 'no one can adequately commend the King's design, which brings such usefulness and advantage to all the inhabitants of Egypt'. His discussion reads:

'For being that the Nile never kept to a certain and constant height in its inundation, and the fruitfulness of the country ever depended upon its just proportion, he dug this lake to receive such water as superfluous, that it might neither immoderately overflow the land, and so cause fens and standing ponds, nor by flowing too little, prejudice the fruits of the earth for want of water. To this end he cut a trench along from the river into the lake, fourscore furlongs in length, and three hundred feet broad; into this he let water of the river sometimes run, and at other times diverted it, and turned it over the fields of the husbandmen, at seasonable times, by means of sluices which he sometimes opened, and at other times shut, but not without great labour and cost; for these sluices could not be opened or shut at a less charge than 50 talents. This lake continues to the benefit of the Egyptians for these purposes to our very day, and is called lake of Myris or Meris to this day.'[21]

Herodotus was mistaken when he considered the Faiyūm oasis (about 50 miles southwest of Cairo) to be the artificial Lake Moeris. The Faiyūm is a natural depression which was once fed by a branch of the Nile, but was separated from the valley before the neolithic age. In the lower part of the depression there was always a natural lake surrounded by marshy areas (figure 5). The pyramid builders, during the third and the fourth dynasties (2600–2450 B.C.), first made an attempt to drain the marshlands and to cultivate those areas. The work continued for several hundred years, and was culminated by the effort of Amenemhet III who reactivated the old

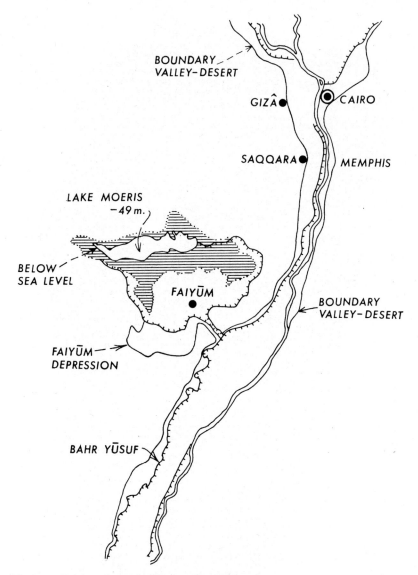

Figure 5. Faiyum depression in the Nile Valley.

branch of the Nile that once led to the Faiyūm. The channel is
known as Bahr Yūsuf or 'Joseph's Arm'.
Herodotus' statement about the circumference of the lake seems to

Figure 6. Records of Nile levels (3000–3500 B.C.) from fragments of an ancient

monument (by courtesy of Palermo Museum, Sicily).

have been much exaggerated as it was highly unlikely to have been more than 110 miles. But the historian's gravest error was about the use of the lake as a flood control reservoir for the Nile – a statement frequently quoted in many books. While it is certain that flood water was used to irrigate the Faiyūm, it is equally certain that none of it could have flowed back to the Nile. The topographical conditions make it completely impossible. Recent investigations[22] have clearly indicated the non-existence of any high level lake in the area during historic times. During the time of Diodorus (30 B.C.) or Strabo (20 A.D.), however, there was some arrangement at Illahun, which was 92 ft above sea level, by which water could either be diverted to the depression or sent back to the Nile by a 50-mile long canal. The device is freely assumed to have been a sluice gate,[21] but it is highly doubtful. Diodorus describes it as a 'skilful and costly device' and Strabo as an 'artificial barrier'. Diodorus also mentions that it could not be *opened or closed* at a less charge than 50 talents' ($ 28,000). The only way these three descriptions can be reconciled is by some system of temporary dams, and that is probably what existed.

NILOMETERS AND FLOOD CONTROL

The day on which the annual inundation of the Nile took place was the most important one in the Egyptian calendar. As the summer season wore on, and the Nile stage gradually increased, people waited anxiously for the *wafa* to come – for *wafa* is the day of celebration and feasting,[23] when the dikes can be cut to allow the life-giving water to flood the land, and thus raise hopes for a good crop for the following year. The levels were marked at several places – notably in a section of the second cataract at Semna. Records of the Nile levels can be traced back to about 3000 to 3500 B.C., from the fragments of an ancient monument now in Palermo Museum, Sicily (figure 6). Nilometers, as the name indicates, were used to measure the levels of the Nile. Markings and inscriptions at several nilometers have been deciphered and correlated. 'At Karnak in 1895, M. LeGrain found a series of 40 high Nile levels marked on the quay walls of the great temple. They date from about 800 B.C., and the mean altitude given by them for a high Nile is 74.25 meters (243.6 ft) above sea-level, while that of today [1906] is 74.93 [probably a misprint 74.25 + 2.68 = 76.93 m], showing a rise of the river bed of 2.68 meters (8.8 ft) in 2800 years, or at the rate of 0.096 meters (3.8 in.) per century.'[24] Jarvis calculated[25] the rate of aggradation of the Nile Delta to be 5.2 in. per 100 years over the period 200 to 1800 A.D.

Three types of nilometers were used. The first type consisted simply of marking the water levels on cliffs on the banks of the river. The second utilized flights of steps which led down to the river. The third and most accurate one used conduits to bring water of the Nile to a well or cistern. The levels were marked either on the walls of the well or on a central column.

The longest continuous record of the Nile is available from the nilometers near Cairo – the most notable one being the Roda (or Rauda) nilometer. When the Arabs conquered Egypt in 641 A.D., they found several *Miqyas an-Nil* (measure of the Nile) in use. The first known Arab nilometer was built on the southern end of the Roda island in 715 A.D., under the Omayyad caliphs Walid ibn 'Abd al-Malik (705–715 A.D.) and Sulaiman (715–717 A.D.). It was rebuilt in 861 A.D. Practically nothing is known about the several other nilometers around Cairo prior to 715 A.D., nor do we know at which particular nilometers the river levels were read.[26] Most of the nilometers now extant, however, originated in the Persian, Ptolemic, and Roman periods.[27]

The Roda nilometer was first investigated by Le Pere and Marcel (1798–1800 A.D.) during the Napoleonic expedition into Egypt. It consisted of a square well connected to the Nile by means of three conduits. At the centre of the well is a graded octagonal column of white marble, and there are steps leading down to the bottom of the well (figure 7). The records of maximum and minimum levels of the Nile are available for the Roda site from 641 A.D. to 1890 A.D. With the completion of Aswan Dam in 1890 A.D., the readings of before and after the date cannot be compared satisfactorily.

The Egyptians depended on the regular inundation of the Nile for their livelihood, but the river in a high flood was entirely another question. It then turned from creation to destruction, and one such flood recorded in 638 B.C. turned the whole valley into 'a primordial ocean, and inert expanse'. Since the Egyptians re-

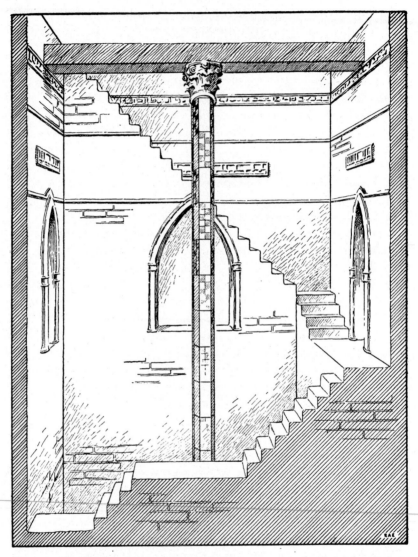

Figure 7. The Roda nilometer in 1798 A.D., a reconstruction from descriptions (by courtesy of University of California Press).

peatedly had to face the problem of serious flooding, they gradually developed a system of flood warning encompassing the various temples and their nilometers. At present, little evidence is available of any actual technical co-operation having taken place between the temples,[28] but according to Diodorus[29] flood warnings were sounded to the population from the nilometer at Memphis in case of emergency. With the approach of the flood season, the levels of the Nile were carefully watched and compared with the markings of the previous year. Swift rowers were sent from the furthest upstream of the gauging stations, one after another, to report the latest level at the capital. These extremely good rowers, rowing with the current, were able to outpace the approaching peak flood and give sufficient advance warning to the townspeople of the forthcoming catastrophe.

Diodorus reports that 'out of the fear of inundation, a watch-tower is built in Memphis, by the Kings of Egypt, where those who are employed to take care of this concern, observing to what height the river rises, send letters from one city to another, acquainting them how many cubits and fingers the river rises, and when it begins to decrease; and so the people, coming to understand the fall of the waters, are freed from their fears, ..., and this observation has been registered from time to time by the Egyptians for many generations'.[28, 29]

The flood observations continued without a break even during major religious or political upheavals. Engreen believes[30] that an interrelation developed between the observations of the Nile flood and the cult of the god Serapis. He quotes Rufinus' statement on the nilometer in the Serapeum at Alexandria in support of his theory: 'But as it was common use in Egypt, that the measuring of the river Nile during its rise was reported to the temple of Serapis, reported, as it were, to the originator of the increased waters and inundations...'.[29, 30]

Various concepts of the origin and rise of the river Nile will be discussed in detail in chapter 6.

UR BABYLONIAN TABLET

The Old Babylonian cuneiform tablet shown in figure 8, presently in the British Museum, presents, on both sides thereof, many problems with their solutions. On the side shown in this figure, 16 problems appear concerning dams, walls, wells, water-clocks, and ex-

Figure 8. Part of Ur Babylonian tablet, 1800 B.C. (by courtesy of Trustees of the British Museum).

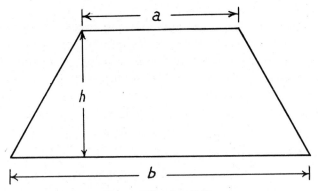

Figure 9. Construction of a dike profile, 1800 B.C.

cavations. There is no information on the precise provenance of this Ur Babylonian tablet (BM 85194), but its date can be said to be roughly in the eighteenth century B.C.

Part of the fourth problem, illustrated by a drawing, requires a dike profile to be constructed in the form of an isosceles trapezoid.[31, 32] The base b, inclination a, and the area A, are given (figure 9), and the top width a is to be calculated. It actually calculates a^2 from

$$a^2 = b^2 - 4aA$$

and

$$A = \left(\frac{b + a}{2} \right) h$$

Hence,

$$4aA = b^2 - a^2$$

In another problem in the same text, a is known, and b is calculated from the first equation.

CODE OF HAMMURABI

King Hammurabi, the self-styled 'obedient and god-fearing prince', conquered Mesopotamia around 1760 B.C. He was the sixth and the greatest King of the first (Amorite) dynasty of Babylon, and was aware of the necessity of having a good network of canals for irrigation as well as for communication and transportation. Unlike the Egyptians with the Nile, the Sumerians had two extremely unpredictable rivers in the Tigris and Euphrates. Flooding was a constant danger, and if it occurred simultaneously in both rivers, it brought untold misery. The story of Noah comes from the legendary greatest flood of the Sumerians,[33] and hence it is no surprise to find the art of building flood protection works, like earthen walls or levees, was quite well developed during this period.

Rulers of this period were as much interested in their hydraulic works as in their conquests, and justifiably so. According to Sarton,[34] traces of their early canals can still be distinguished from the air. The available documentary evidence indicates that King Hammurabi often directed his provincial governers to dig canals as well as to dredge them regularly. In a memorandum to one Sid-idinnam, the King complained about a canal having been so imperfectly dug that boats could not enter Erech, and also about the canal needing repair near the banks of the Druru. He ordered an official to rectify those defects within three days of receipt of his communication, using men who were then at his disposal. Thus, with the gradual political integration of the country, there was a strong centralized control of water.

The famous Code of Hammurabi is the most complete codification of Sumerian and Babylonian law. It was discovered in Susa in 1901 by the French Assyriologist Jean Vincent Scheil, and is at present in the Louvre Museum in Paris. Figure 10 shows the upper part of the Code of Hammurabi, depicting the King as either offering it to the Sun God Shamash or being charged by the God to write his code. The laws concerning irrigation were carefully contrived, and it seems that they were primarily aimed at preventing carelessness which might result in flood damages, as emphasized in the following excerpts:

'Sec. 53. If any one be too lazy to keep his dam in proper condition, and does not keep it so; if then the dam breaks and all the fields are flooded, then shall he in whose dam the break occurred be sold for money and the money shall replace the corn which he has caused to be ruined.'

'Sec. 55. If any one open his ditches to water his crop, but is careless, and the

Figure 10. Part of the code of Hammurabi in the Louvre Museum.

water flood the field of his neighbor, then he shall repay his neighbor with corn for his loss.'
'Sec. 56. If a man let out the water, and the water overflow the land of his neighbor, he shall pay 10 *gur* of corn for every 10 *gan* of land flooded.'[35]

In the absence of further records, it can safely be assumed that hydrologic engineering was quite advanced nearly 4000 years ago, and that even today one can detect the influence of the Code of Hammurabi on modern water laws.

SINNŌRS OF PALESTINE

Sinnōrs, or water tunnels, were used in Palestine prior to 1200 B.C. The cities in Palestine and Syria were usually built on the tops of hills at the bottoms of which were streams providing the municipal water supplies. Thus, during times of war, cities were rather vulnerable because invaders could easily cut off the supply of water from the city. To protect the city, first a tunnel was dug, one end of which provided a secret approach to the stream. Its other end was located within the city's boundary. Entrance to the sinnōr was gained by a shaft provided with a flight of stairs. In later versions thereof, a conduit on the floor of the tunnel brought water from the stream to the base of the shaft.[36]

Figure 11 shows the Siloam sinnōr which King Hezekiah constructed around 700 B.C. According to the Second Book of Chronicles, the King also 'stopped the upper watercourse of Gihon, and

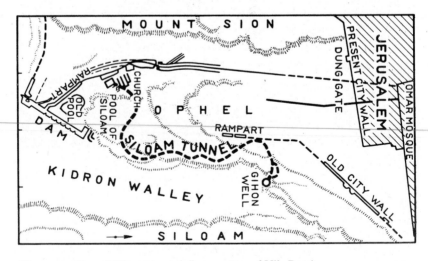

Figure 11. Plan of Siloam tunnel (by courtesy of Nils Borg).

brought it straight down to the east side of the city of David'. The 1750 ft long tunnel, cut through limestone, conducted the water of the Gihon well under the city wall into the city proper. The tunnel is still in use today.

EVIDENCE AT NIPPUR

Around 1300 B.C., Nippur in Mesopotamia was one of the most famous centres of Sumerian religion, and recently American archaeologists have excavated thousands of tablets at the site. Most of these tablets were unbaked, and hence they are not well preserved, and are very difficult to decipher, but some of the maps excavated at Nippur were so faithfully made that they even helped the archaeologists to continue their excavations.

Figure 12 is a map of fields and canals at Nippur. A translation of the cuneiform captions is shown in figure 13. Unmarked fields were either common lands for grazing or in the process of changing hands. Probably the 'Canal of Hamri' was the main canal, and water entered it during periods of the river's high flow. It is likely that channels marked 'irrigation' or 'canal' were intended for irrigation purposes, whereas those marked 'stream' could have been drainage ditches. The drainage ditches were perhaps used to remove the water out of cultivated areas during low-river stages. It is interesting to note the designation 'Marshland of the town of Hamri', where in all probability cane-reed was cultivated. Since the country was almost treeless, reeds were used for almost everything – building, basketry, furniture, and firewood. Reed-matting was used for strengthening the dikes which protected the alluvial soil.

WATER METERS

A primitive type of water meter was used at the Gadames oasis of North Africa more than 3000 years ago,[37-39] and it is still being used without modification. The oasis has a small spring called Ain el Fras ('Spring of the Mare') which according to legend, was discovered by the horse of an Arabian conqueror. The spring discharges around 180 cu. m per hr, which is collected in a basin and distributed through a main canal and two side canals. The

Figure 12. Map of irrigation system near Nippur in Mesopotamia, around 1300 B.C. (by courtesy of University of Pennsylvania Museum).

process was developed for the equitable distribution of irrigation water from the spring to the various agricultural fields.

A container, consisting of a pot with a hole in its bottom, is lowered by means of a rope into the water of a well at the market place. When filled it is pulled up, the water is allowed to drain back into the well through the hole in the pot, and the cycle is repeated over

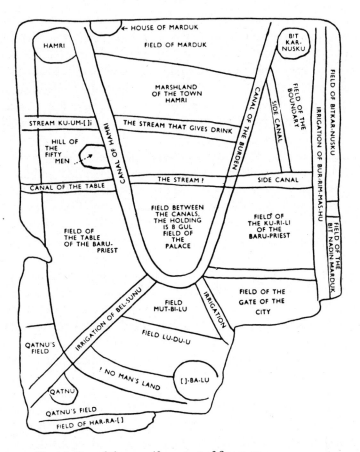

Figure 13. Translation of the cuneiform text of figure 12.

and over again. It requires a period of about three minutes for each cycle to be completed. This arrangement forms the basis of the water-distribution system. A water commissioner decides the amount of time, as measured continuously by the filling and emptying of the pot, that a given land owner may divert the entire flow of the nearest irrigation ditch onto his land. Upon the arrival of the last cycle to which a given participant is entitled, a small bundle of straw is dropped into a nearby reach of the irrigation ditch, and allowed to float down to that man's field. When it arrives there, he must clear the channel so as to allow the water to flow down to the dam of the next participant, and he must also seal off his own diversion sluices. That process is repeated day after day and night after night until the last participant along the ditch has received all of the water released to him while his quota of cycles are being counted off at the well in the market place. After that last field has been supplied with its quota of water, the entire process is repeated, beginning with the field nearest the supply reservoir. About 12 days are required to make the complete round. Undoubtedly it is a crude way to distribute such flowing water, but the fact that the process is still being carried out without any appreciable changes, is evidence that it is reasonably effective. It is amazing that day and night for over 3000 years there has always been a man in attendance at that well to see that the water from the spring is equitably distributed.

Another type of water meter, whose principle of operation is just the reverse of the one just described, has been in continuous use in Yemen for many years.[40] In this instance, a 'floating clock' is used to time the distribution of water from a small stream. An empty copper bowl is placed in a larger copper container that is full of water. The bowl gradually becomes filled by the water entering through a hole in its bottom, and it sinks in about 5 minutes. It is then immediately raised to the surface to repeat the unending process. Each farmer is permitted in his turn, to divert the entire flow of the stream onto his land for a certain period of time as measured by the 'floating clock'. That period is computed on the basis of one cycle of 'bowl-filling' time for each 100 lubnah (approximately $2/3$ acre) of land in the field. Because of the importance of water in the desert, the job of Muqassim addayri (the water-divider) is important and prestigious.

GROUND WATER UTILIZATION

Undoubtedly the greatest achievement in the utilization of ground water of ancient times was the building of qanāts (or kanāts). A qanāt is an artificial underground channel which carries water over long distances either from a spring or from water-bearing strata, and it solved several problems in water resources engineering. First, evaporation was undoubtedly a major problem in hot and arid climates, and hence, with the limited water supply, surface transport

Figure 14. Aerial photographs of qanāt systems in Persia.

was a distinct hazard. Secondly, it was difficult to maintain a uniform slope in a hilly country; and finally, qanāts kept water cool
and free of surface pollutants. Figure 14 is an aerial photograph of
qanāt systems originating in the talus deposits at the foot of the
mountain near Kashan in Persia.

Contrary to present belief, qanāt building probably started in
Armenia[36, 41] and not in Persia. In his invasion of Urartu (present
Armenia), King Saragon II (721–705 B.C.) of Assyria destroyed
the irrigation network of the town of Ulhu. He described the irrigation system of the vanquished King of Ulhu in these terms:
'Following his ingenious inspiration Ursa, their King and Lord ...
revealed the water outlets. He dug a main duct which carried
flowing waters ... waters of abundance he caused to flow like the
Euphrates. Countless ditches he led out from its interior ... and he
irrigated the fields'.[42]

The construction of this remarkable system, which according to
Tolman[43] was 'the greatest waterworks of the ancients', was
directed by an engineer called *Muqannī*. He first located the water-
bearing strata by digging a number of test wells, and when a good

stratum was hit, a mother well was dug. Another well was dug some distance away, usually about equal to the depth of the well, and the two wells were connected by a tunnel. By this procedure the construction continued. The direction and depth of the tunnel was determined by means of a crude but adequate system of plumb bobs. Figure 15 shows a typical water supply system by qanāts; the cross-section was somewhat egg-shaped. Since the Persians rarely

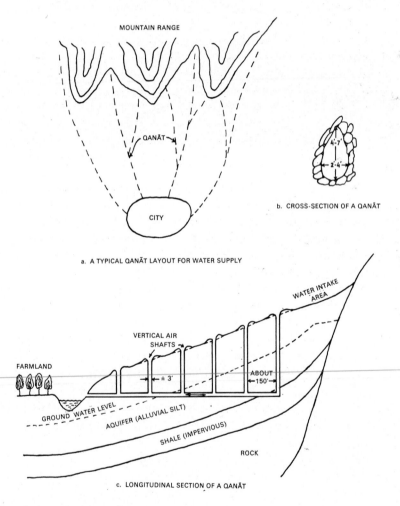

Figure 15. Details of qanāt system (not to scale).

dug through rock, the routes of the qanāts had numerous twists and turns, and large deviations were also made around hills. Only one man could dig at one time, and the excavated material was removed in a goat-skin bag through vertical air shafts. If necessary (depending on the soil conditions), lining materials were carried in the same bag on the return journey. As reflected light was used for digging, and working conditions were rather grim, one would expect that accidents would be rather common and loss of lives frequent. Details of qanāt construction have been discussed in detail by Wulff.[44]

South of Dizful in Persia is one of the old qanāt systems. It consists of three pairs of tunnels taking water from the gravel bars near the River Ab-i-diz, about seven miles north of Dizful. Two pairs of qanāts supply water to the neighbouring land for agriculture, and the remaining pair supplies the city. These qanāts are at such a depth that some houses in the city extend six stories below the ground level to tap the water.

During the time of Darius I (521–485 B.C.), his Caryandan Admiral Scylox went to the oasis of El Khargeh[45] in Egypt, and there introduced the qanāt system of irrigation. Butler believed[35] that they must have extended far enough eastward – in fact under more than a hundred miles of rolling desert – to intercept seepage from the Nile. Recent investigations, however, have clearly indicated[46] that the trace of the qanāts can be found from discharge point back toward the intersection of the water table in the talus slope of the escarpment of the plateau, a distance of about 2 miles. Traces of it can be followed very easily by the vertical air shafts connected with the main ditch.

Use of these long infiltration galleries to tap ground water from soft sedimentary rocks or alluvial fan deposits quickly spread from Armenia to as far as Northern India.

The qanāt used the principle of gravity flow. Its average length in desert regions was 25 to 28 miles. It had a gentle slope of 1 to 3 on 100. In some places it had a depth of nearly 400 ft.[47] Considering the state of hydraulic science during the period in question, it was no mean achievement.

MARIB DAM

The Marib dam, which was probably constructed somewhere between 1000 and 700 B.C., was considered to be one of the wonders of
the ancient world.[48] It was located on the Wadi Dhana, nearly
40 miles from the ancient city of Marib (now in Yemen), and was
known to the Moslems as Sudd-el-Arim. According to the Encyclopedia of Islam,[49] a series of dams controlled the River Denne, a
fair-sized river on the eastern side of the high mountain range in
Yemen.[18] Of all the dams on the river, Marib was the largest and
was an earthfill structure. The dam was some 33 ft high and 1900 ft
long. It was flanked on either side by large outlet works of excellent
masonry.[50] No mortar was used in the construction of the dam (just
as in the ill-fated dam at Sadd el-Kafara), except for a covering
on the top, which was probably added for the prevention of
damage from rain.
The dam first breached in the fifth century A.D., and was finally
destroyed during the latter half of the next century. But unlike
Sadd el-Kafara, its failure had terrible consequences. 'There is
hardly any historic event of pre-Islamic history that has become
embellished with so much that is fanciful, and related in so many
versions, as the history of the bursting of the Marib Dam.'[49] According to the Koran (Sura 34, verse 14) 'the people of Seba had
beautiful gardens with good fruit. Then the people turned away
from God, and to punish them, He burst the dam, turning the good
gardens into gardens bearing bitter fruit'.

WORKS OF SENNACHERIB

No work on ancient hydrologic engineering can be complete without
reference to Sennacherib (705–681 B.C.). Out of shear barbarism,
the Assyrian King dammed the Euphrates in 689 B.C., only to let
loose the resulting flood on the vanquished and burnt city of
Babylon. But Sennacherib was an excellent strategist, a good
architect, and among other things he invented water-lifting machines
and introduced cotton to Assyria. His main contribution to hydrology, however, was his successful attempt to develop the water
resources of the region.[50]

The capital of Assyria was Nineveh, on the banks of the River Tigris, but the muddy water of the river was not good enough for an emperor, and hence he sought clear water for Nineveh and his palace at Khorsabad. It was accomplished in three stages. The first stage involved placing a weir across the River Khosr near Kisiri (703 B.C.), and constructing a 10-miles-long canal to irrigate the orchards of some 18 settlements in the plain west of Nineveh (figure 16). In the second stage (694 B.C.) 18 water courses were diverted and canalized to bring water from the mountainous Jebel

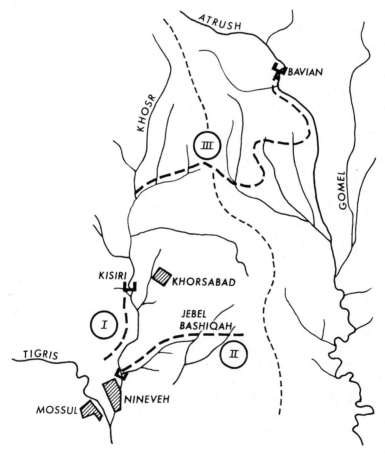

Figure 16. Water resources development of King Sennacherib.

Bashiqah region to the city. Two dams, constructed of square stone blocks, were used in series to divert this water from the Khosr River to the capital. The diversion of water from the Atrush-Gomel River system (690 B.C.) was accomplished during the final stage. To do so, a diversion dam was built obliquely across the Atrush River near the gorge of Bavian to create a reservoir. Then he constructed a magnificent canal[51-55] (Sennacherib's Channel), for conducting the water from the reservoir. The canal followed the natural course through the foothills with a slope of about 1 on 80. It took 15 months to complete this 35-mile long canal with its extensive limestone-block pavements and its arched aqueducts over several brooks and valleys. The aqueduct of Jerwan was nearly 1000 ft long, 39 ft wide, with two side walls 5 ft high and 8 ft thick. Fourteen piers were used to carry the aqueduct, the bed of which consisted of three layers of limestone blocks. The inscription on the structure stated, 'Over deep-cut ravines I spanned (lit., caused to step) a bridge of white stone blocks. These water I caused to pass over upon it'.[44] A rock carving near the headwork of the structure records, 'To the great gods the King prayed and they heard his prayer; they directed the work of his hands. By gates and a tunnel the sluices opened of themselves and permitted the rich water to flow down ... According to the wishes of the Gods' heart, he had dug the water, carried from the stream, and directed its force'.

CONCLUSION

The various projects discussed in this chapter are just a few of the more outstanding hydrologic engineering works of the ancient times. It must be realized that they formed only a minor part in the man's overall struggle towards civilization. The huge tasks that had been successfully undertaken by countless of nameless people were major operations even when judged by the present-day standards. The marshy and unhealthy flood-plains of the Nile, Tigris-Euphrates, Indus and Huang Ho had to be reclaimed, levees had to be made for flood protection, numerous wells and canals had to be dug, and systems of irrigation and drainage had to be inaugurated. The whole process was obviously so successful that even the Garden of Eden was claimed to have been located between the

Tigris and the Euphrates rivers. Hydrologists and hydraulic engineers played a significant part in the successful planning, design, construction, operation and maintenance of these works. Their works were instrumental in the development of these civilizations – so much so that we can justifiably call them 'hydraulic civilizations'.

REFERENCES

1. BISWAS, ASIT K., Hydrology during the Hellenic Civilization. Bulletin, International Association of Scientific Hydrology *12* (1967) 5–14.
2. EASTON, S. C., The heritage of the past, revised ed. New York, Holt, Rinehart and Winston Inc. (1965) p. 35.
3. DALES, G. F., The decline of the Harappans. Scientific American *214* (1966) 92–100.
4. BISWAS, ASIT K., A short history of hydrology. Proceedings of the International Seminar for Hydrology Professors, University of Illinois, Urbana (1969).
5. DROWER, M. S., Water-supply, irrigation, and agriculture. In: A history of technology, edited by Charles Singer, E. J. Holmyard and A. R. Hall, vol. 1. London, Oxford University Press (1954) pp. 520–557.
6. LANE, E. W., Manners and customs of the modern Egyptians, edited by E. Rhys. London, Everyman's Library, J. M. Dent and Sons Ltd. (1908) ch. 14.
7. HERODOTUS, The history of Herodotus, translated by George Rawlinson. Great Books of the Western World, vol. 6, book 2. Chicago, Encyclopaedia Brittanica Inc. (1952) pp. 68–69.
8. SCHNITTER, N. J., A short history of dam engineering. Water Power *19* (1967) 142–148. Discussions by Asit K. Biswas, *19* (1967) 258. Reply by N. J. Schnitter, *19* (1967) 345.
9. SCHWEINFURTH, G. A., Auf unbetretenen Wegen in Aegypten. Hamburg, Hoffman und Campe (1922) pp. 213–231.
10. MURRAY, G. W., Water from the desert: some ancient Egyptian achievements. The Geographical Journal *121* (1955) 171–181.
11. MURRAY, G. W., Desiccation in Egypt. Bulletin de la Société Royale de Géographie d'Egypte *23* (1949).
12. HELSTRÖM, B., The oldest dam in the world. Bulletin no. 28, Institution of Hydraulics, Royal Institute of Technology, Stockholm (1951).
13. DUSSAUD, R., La digue du lac de Homs et le 'mur Egyptien' de Strabon. Monuments et Mémoires (1921–22) 133–141.
14. MATSCHOSS, C., Great engineers, translated by H. S. Hatfield. London, G. Bell and Sons Ltd. (1939) p. 5.
15. NEEDHAM, J., Science and civilisation in China, vol. 3. Cambridge, University Press (1959) p. 515.
16. BRITTAIN, R. E., Rivers, man and myths, from fish spears to water mills. New York, Doubleday and Co. Inc. (1958) p. 59.

17. PAYNE, R., The canal builders. New York, The Macmillan Co. (1959) pp. 13–16.

18. HATHAWAY, G. A., Dams – their effect on some ancient civilizations. Civil Engineering, ASCE *28* (1958) 58–63.

19. HERODOTUS, *op. cit.*, p. 81.

20. WILLCOCKS, W., From the garden of Eden to the crossing of the Jordan, 2nd ed. London, E. and F. N. Spon Ltd. (1926).

21. DIODORUS SICULUS, The historical library of Diodorus, the Sicilian, translated by G. Booth, vol. 1. London, J. Davis (1814) p. 57.

22. BISWAS, ASIT K., Hydrologic engineering prior to 600 B.C. Journal of Hydraulics Division, ASCE *93* (1967) 115–135. Discussions by G. Garbrecht, G. J. Requardt and N. J. Schnitter, *94* (1968) 612–618, by B. F. Snow, *94* (1968) 805–807, and by F. L. Hotes, *94* (1968) 1356–1357.

23. THOMSON, M. T., The historic role of the rivers of Georgia: measurements from Mena to Mead. Privately circulated.

24. LYONS, H. G., Physiography of the River Nile and its basin. Cairo, Ministry of Finance, Survey Department (1906).

25. JARVIS, C. S., Flood-stage records of the River Nile. Proceedings, ASCE *62* (1936) 1012–1071.

26. POPPER, W., The Cairo nilometer. Publications in Semitic Philology, vol. 12. Berkeley, University of California Press (1951).

27. BORCHARDT, L., Nilmesser und Nilstandsmarken. Berlin, Preussische Akademie der Wissenschaften, Philosophisch-historische Abhandlungen nicht zur Akademie gehöriger Gelehrter, no. 1 (1906).

28. OTTO, W. G. A., Priester und Tempel im hellenistischen Aegypten. Leipzig, B. G. Teubner, vol. 1 (1905) pp. 22, 43–44, vol 2. (1908) pp. 311–313.

29. DIODORUS SICULUS, *op. cit.*, p. 42.

30. ENGREEN, F. E., The nilometer in the Serapeum at Alexandria. Medievalia et Humanistica *1* (1943) 3–13.

31. WAERDEN, B. L. V. D., Science awakening, translated by A. Dresden. Groningen, P. Noordhoff Ltd. (1954) p. 68.

32. NEUGEBAUER, O. and A. SACHS, JR., Mathematical cuneiform texts. New Haven, Conn., American Oriental Society and American Schools of Oriental Research (1945) pp. 96–97.

33. KAZMANN, R. G., Modern hydrology. New York, Harper and Row Inc. (1965) pp. 1–20.

34. SARTON, G. A., History of science: ancient science through the Golden Age of Greece. Cambridge, Mass., Harvard University Press (1959) p. 79.

35. BUTLER, M. A., Irrigation in Persia by kanats. Civil Engineering, ASCE *3* (1933) 69–73.

36. FORBES, R. J., Studies in ancient technology, vol. 1. Leiden, E. J. Brill (1955) p. 151.

37. RICHARDSON, C. G., The measurement of flowing water. Water and Sewage Works *102* (1955) 379–385.

38. SCHAAK, M., Hundert Jahre Wassermessung. Neue Deliwa Zeitschrift zur

Förderung des Gas-, Wasser- und Elektrizitätsfaches, Hanover *1* (1953) 132–135.

39. COMMITEE 861D, Water meters – Selection, installation, testing, and maintenance. Chapter 1, Early history of water measurement and the development of meters. Journal, American Water Works Association *51* (1959) 791–799.

40. ABERCROMBIE, T. J., Behind the veil of troubled Yemen. National Geographic Magazine *125* (1964) 423.

41. LEHMANN-HAUPT, C. F., Armenien einst und jetzt, vol. II. Berlin, B. Behr (1910) p. 111.

42. LASSØE, J., The irrigation system at Ulhu. Journal of Cuneiform Studies *5* (1951) 21–32.

43. TOLMAN, C. F., Ground water. New York, McGraw-Hill Book Co. Inc., (1937) pp. 1–25.

44. WULFF, H. E., The qanāts of Iran. Scientific American *218* (1968) 94–105.

45. CATON-THOMPSON, G. and E. W. GARDNER, The prehistoric geography of Kharga Oasis. Geographical Journal *80* (1932) 369–409.

46. LAMOREAUX, P. E., Personal communication (1968).

47. FEILBERG, C. G., Qanaterne, Irans underjordiske vandingskanlen. Copenhagen, Øst og Vest (1945) pp. 105–113.

48. PHILLIPS, W., Qataban and Sheba: exploring the ancient kingdoms on the biblical spice routes of Arabia. London, Victor Gollanc Ltd. (1955) pp. 200–201.

49. ENCYCLOPEDIA OF ISLAM, vol. 3, Leiden, E. J. Brill (1911–38) pp. 263–293.

50. GLASER, E., Reise nach Marib. Vienna, Ed. A. Hölder (1913).

51. BROMEHEAD, C. E. N., The early history of water supply, part II. Geographical Journal *99* (1942) 183–193.

52. FINCH, J. K., Master builders of Mesopotamia. Civil Engineering, ASCE *27* (1957) 50–53.

53. MANER, A. W., Public works in ancient Mesopotamia. Civil Engineering, ASCE *36* (1966) 50–51.

54. JACOBSEN, T. and S. LLOYD, Sennacherib's aqueduct at Jerwan. Oriental Institute Publications, vol. 24. Chicago, University of Chicago Press (1935).

55. SCHMÖKEL, H., Ur, Assur und Babylon. Stuttgart, Killper Verlag (1955).

2

The Hellenic Civilization

INTRODUCTION

The pre-Hellenic civilizations, as discussed in the previous chapter, grew up mainly on the banks of the three major river systems: the Nile, the Tigris–Euphrates, the Indus, and their tributaries. The Hellenic Civilization came into being in about 600 B.C. with the birth of the Ionian school in Asia Minor. Its science was certainly indebted to older civilizations, mostly perhaps to that of the Egyptians, but there, for the first time, people engaged in a pursuit of knowledge for its own sake. Reymond said of it: 'Compared with the empirical and fragmentary knowledge which the peoples of the East had laboriously gathered together during long centuries, Greek science constitutes a veritable miracle. Here the human mind for the first time conceived the possibility of establishing a limited number of principles, and of deducing from those a number of truths which are their rigorous consequence'.[1]

The failure of the Greek philosophers to establish some of the basic principles of water science was not due to any neglect of facts, as they certainly had noted certain aspects of those principles, but the number thereof was too few. Moreover, they failed to undertake a consideration of all the facts in their possession simultaneously. Instead, they followed a procedure, as recommended by Aristotle, to consider only small portions of them at a time. It is not surprising therefore that many conflicting theories were advanced on almost every subject. This circumstance was summed up by Rapin as follows:

> 'All the power of ancient philosophy was not able to settle any one principle of nature. Thales maintain'd that the water was the great source of all things; Heraclitus declar'd for the fire; Anaximenes for the air; Pythagoras for numbers; Democritus for atoms; Museus for unity; Parmenides for infinity. . . Protagoras affirmed that every thing was really true which appear'd to be so. Aristippus allow'd nothing to be true but what men are thro'ly convinc'd of by inward perswasion of the mind. Chrysippus declares, that the senses are always in the wrong: Lucretius contends, that they are always in the right.'[2]

Rapin concluded charmingly that 'it must be confess'd that there's nothing so certain in nature, but what·may be made the subject of dispute'.[2]

THALES, THE ANCIENT HYDROLOGIST

At the threshold of the Ionian philosophy, stood the semi-legendary yet very real and outstanding figure of Thales of Miletos (624?–548? B.C.). Very little is known about him (figure 1). Whatever we are able to credit him with comes mostly from Aristotle and Herodotus. Even during the fourth century B.C., the time of Aristotle, Thales was well on his way toward becoming a legendary figure. Sarton speaks of Thales as a sort of Benjamin Franklin. Both of those men had open minds, a curiosity to learn about new elements, and both applied their knowledge to the solution of practical problems. Like Franklin's visit to England, Thales went to Egypt, and like Franklin, he was impressed with what he saw.

Thales was acknowledged universally as one of the Seven Wise Men of the ancient times (figure 2). It is amusing to note the statement by Demetrios Phalersus that Thales 'received the title of sage' – as if it were a sort of honorary doctorate! The unusual fame of Thales and the esteemed title of 'wise man' did not come to him from the fact that he was the first Greek astronomer nor because he was the first Greek mathematician, but rather because of the application of his knowledge to practical advantage.[3]

In a dissertation on the history of hydrology, we are interested in the following statements which can, according to Aristotle, be attributed to Thales:

Figure 1. Thales, the founder-member of the Seven Wise Men (by courtesy of Ny Carlsberg Glyptotek, Copenhagen).

(1) The earth floats on the water,[4] and
(2) Water is the original substance, and hence is the material cause of all things.[5]

Figure 2. Mosaic of the Seven Sages from Torre Annunziata near Pompeii. Identification is rather difficult, but the third from left is probably Thales (by courtesy of National Museum, Naples).

The first of these two statements was quite common in the Egyptian and the Babylonian cosmogonies. The Egyptian priests believed that the earth was created out of the primordial waters of Nūn and that such waters were still everywhere below it.[6] According to a Babylonian legend prevailing at that time, 'All the lands were sea ... Marduk bound a rush mat upon the face of the waters, he made dirt and piled it beside the rush mat'.[7] Marduk was the legendary Creator of the Babylonians, and since Thales travelled extensively, it is reasonable to assume that he knew of that legend, but when he expressed the above theory, Marduk was not mentioned. It also seems possible that Thales might have obtained his theory from Homer who thought that the earth was surrounded by a vast expanse of water beyond the sea – which has no source or origin. Or Thales possibly tried to rationalize the Egyptian and the Babylonian theories, and ended up with the concept that the earth floated on water like a flat disc.

His second statement that water is the fundamental or primary matter might sound foolish at first reading but appears quite reasonable after closer scrutiny. It is far from being nonsense. When Thales visited Egypt, he must have observed the Nile – an almost legendary river. It was then a common belief that Egypt had been created by the Nile. And, if he had noticed the process of delta building by the river, he might well have reasoned that the land was actually being produced by the water. The very lives of the Egyptians depended entirely on the regular inundations of the Nile, yet they never seem to have presented a rational reason or explanation for them. Eventually, the Ionian offered an explanation for the phenomenon, but to the Egyptians this was just another of their numerous mysteries.

Water is the only liquid that was well known in its three physical states – liquid, vapour and solid. Its transformation from one state to another takes place quite easily, and it is commonly found in nature in all three states. Because it comes down from above as rain, hail or dew, some early philosophers erroneously believed that it became changed into earth by some mysterious process. The thought of associating the origin of streams and rivers with precipitation was at first an extremely remote possibility – a concept that will be discussed in later chapters. Another erroneous belief was that the fires of heavenly bodies were fed by the moisture drawn from rivers and seas through the process of evaporation.[8] Probably a major factor which caused Thales to assume that water was the primary element, was that life cannot exist without it. That fact must have been driven home rather forcefully when he visited the arid country of Egypt.

Some authorities like Oswiecimski have tried to justify Thales' concept on somewhat different grounds. He contended that:

'essence of water is, of course, what is most obvious in water, i.e., humidity or
fluidity which can be easily identified with humidity. It was easy for Thales to
observe that such a solid and 'dry' body in its normal state as iron even in its
liquid condition (e.g. when melting ore) is not only water, but even, under the
influence of heat, cannot contain any water in the usual sense of the word. Yet
it is a liquid, and liquid in general is easy to be identified with humidity. So if
iron is really, according to Thales' theory, one of the forms of water or – what
is the same – arose from primitive water, it is not the form of water itself but
of its essential part, of its essence: humidity–fluidity which he understood in a
material sense, not as a quality but rather as a concrete matter.'[9]

Situated at the dispassionate distance of some 2500 years, it is
futile to argue against the actual contention of the founder of the
Milesian school. A woman wearing her grandmother's dress may
look like her grandmother but her thoughts, no doubt, would be
different. Hence we refer to Aristotle to obtain an insight into
Thales' reasons for describing water as the primary substance.
Aristotle suggested three reasons for Thales' philosophy: (a) nutri-
ment of all things is moist, (b) heat is generated from moisture and
also kept alive by it, and (c) seeds of all things have a moist con-
stituency.

Like all ancient philosophers Thales was fascinated by the River
Nile. His attempts to explain the cause[10-14] of its regular inundations will be discussed in chapter 6.

The search for the primary substance or *archē* continued long after
Thales' time. Heraclitus thought it was fire, and the priests of magic
broadened the concept to include both fire and water. Euripides
considered the primary substances to be air and earth, and believed
that the generation of mankind and the animals was due to the fact
that the earth was impregnated by the seeds contained in the
precipitation[15] from the heaven. When all the living things were
destroyed by time, they returned again to their point of origin – the
heaven. Empedocles of Agrigentum (490–430 B.C.) postulated that
there were four primary elements or roots (rhizōmata) – fire, air,
water, and earth, from which all the materials of the world were
constituted by their combination in different proportions.[16] Bones,
for example, were made of two parts of earth, two of water, and
four of fire. This concept of the constitution of materials by different
ratios of the four 'elements' probably came about through the
mathematical influence of Pythagoras.

ANAXIMANDER TO XENOPHANES

Anaximander of Miletos (610–545 B.C.) was a contemporary of
Thales, and hence it is not surprising to find that Thales influenced
him to a certain extent. He considered the Thalesian concept of
water as the primary substance to be too tangible, and hence he
selected something more intangible which he called *apeiron*. That
word meant something infinite, indefinite, undetermined or even
inexperienced. Since only fragments of Anaximander's works are
available to the present historians of science, there is considerable
controversy over the true meaning of the word *apeiron*.[17] Like
Thales, Anaximander believed that in the beginning, human, as
well as animal life, originated in water. With gradual but continuous
evaporation, land emerged where once was an all-engulfing sea.[18]
Aristotle later discussed that view as follows: 'But those who are
wiser in the wisdom of men give an [explanation of the] origin of
the sea. At first, they say, all the terrestrial region was moist; and,
as it became dried up by the sun, the portion of it that evaporated
produced the winds and the turnings of the sun and the moon,
while the portion left behind was the sea. So they think the sea is
becoming smallest by being dried up, and that finally it will all
become dry.'[19]
According to Hippolytos, Anaximander believed that precipitation
was due to the moisture being drawn up from the earth by the sun.[20]
Anaximenes (d. 528–525 B.C.) believed that when rain is frozen
while falling it resulted in hail, whereas snow was produced when
air was imprisoned within the water.[21] Xenophanes of Colophon
lived somewhere within the period 570 to 470 B.C. He believed
that the 'sea is the source of water, and the source of wind. For
neither could (the force of the wind blowing outwards from within)
come into being without the great main (sea), nor the stream of
rivers, nor the showery water of the sky; but the mighty main is
the begetter of clouds and winds and rivers'.[22] Thus Xenophanes
presented an argument of purely tautological character to prove
his point. Clouds, rains, springs, and streams, he claimed, all
originate from the sea;[23] if there was no sea, none of these would
have existed but since there is sea, they do exist! From his observation of the presence of shells on high mountains and fossils of

marine animals at various land areas of the earth, he reasoned that the land must have been under the sea at one time. For those observations, he could be considered the earliest geologist as well as the earliest palaeontologist.[24]

ANAXAGORAS AND HIPPON

Anaxagoras of Clazomenae (500–428 B.C.) was endowed with a spirit of inquiry, and was the last of the renowned Ionian philosophers. His explanation for the regular rise of the river Nile was almost the correct one, and it will be discussed in chapter 6. His main thoughts concerning hydrology were the following:

'Of the moisture on the surface of the earth, the sea was produced from the waters of the earth, . . . and from the rivers which flow into it.'
'Rivers depend for their existance on the rains and on the waters within the earth, as the earth is hollow, and has water in its cavities. And the Nile rises in summer owing to the water that comes down from snows in Ethiopia.'[25]

Hippon of Samos flourished around the middle of the fifth century B.C. According to him: 'All water that is drunk comes from the sea; for of course the wells from which we drink are not deeper than the sea, for in that case the water would not be from the sea but from elsewhere. But in fact the sea is deeper than the water. It follows therefore that all water that is above the sea comes from the sea'.[26]

CONTRIBUTIONS OF HERODOTUS

Herodotus of Halicarnassus (484–425 B.C.) considered all knowledge to be his particular prerogative, and pursued that subject with great enthusiasm. Hydrologic phenomena were included among the many things he was particularly curious about and so he searched diligently for their explanations. Any reasons he could find, rational or irrational, were carefully entered in his notebook. As an example, he described the three prevailing theories as to why the River Nile's inundation began at the commencement of the summer solstice. Those three theories had all been expressed by the Greeks, and that fact caused him to note that the Egyptians did not have any theories thereon. Regardless of their source, however, he dismissed them all disdainfully, and submitted a fantastic theory thereon of his own accord. More information about it will appear in chapter 6.

Herodotus studied the Nile with particular interest. He said that any one with only ordinary powers of observation could see that Egypt was an acquired country, a gift of the river. Its alluvial land had been gradually built up by the deposition of silt brought there by the river.[27] If one dropped a sounding line a day's sailing time away from the coast, he would find mud there at a depth of eleven fathoms.[28] That indicated that soil eroded by the river had been carried for that distance. With brilliant geological reasoning almost unparalleled in ancient history, Herodotus opined that all of Lower Egypt had once been under the sea. Like the Red Sea, the Nile valley was once an arm of the sea but silt carried by the river gradually filled in the basin between Thebes and Memphis.[29] The delta formed while the area was gradually being filled. That filling had taken place during the 'ages that passed before I was born by the great River Nile which works great changes'.[30] The presence of sea shells on the hills and the high salinity of the land, helped to confirm his conclusions.[31]

The Egyptians became amused as well as horrified when they learned that Greece did not have a river such as the Nile for producing an annual inundation, and that rainfall was the only source of fresh water in Greece. They believed that if the Gods decided not to send rain to Greece, the poor Greeks would become wretchedly hungry – a belief which must have caused the father of history to chuckle to himself. According to 'strong evidence' provided by the priests at Heliopolis, the Nile, during the reign of King Moeris, overflowed all Egypt below Memphis as soon as the river stage rose to only eight cubits (12 ft). When Herodotus visited Egypt some 900 years later, however, the river had to rise to sixteen cubits (24 ft) to produce the same effect.[32] If the land kept on increasing in height at the same rate, it was obvious that a time would come when the Nile would no longer be capable of flooding its banks. Without an annual inundation in an almost rainless country, this astute historian reasoned, the chances were that it would be the Egyptians who would then go hungry, rather than the Greeks.

Herodotus was fascinated by the Ister River (Danube), almost as

much as he was by the Nile. In contrast to the Nile, which overflowed its banks with almost unbelievable regularity, the Ister maintained practically the same level during the summer as it did during
the winter.[33] This was true despite the fact that it snowed during
the winter and there was scarcely any rain, whereas during the
summer extra water was brought to the river by both melting snow
and rainfall. The sun's power of attraction was greater during the
summer, however, and because the two effects tended to counter-
balance each other, the flow in the Ister remained at essentially the
same level throughout the entire year.[34] Where did the Ister get its
supply of water during the winter to maintain that flow? The
historian has not told us, and it is futile to speculate as to what
reason or reasons he could have given if he were to explain that
circumstance. Whewell believed that Herodotus' statement that
'the sun *draws*, or attracts, the water' was a metaphorical term,
'obviously intended to denote some more general and abstract
conception than that of the visible operation which the word
primarily signifies. This abstract notion of 'drawing' is, in the
historian, as we see, very vague and loose; it might, with equal

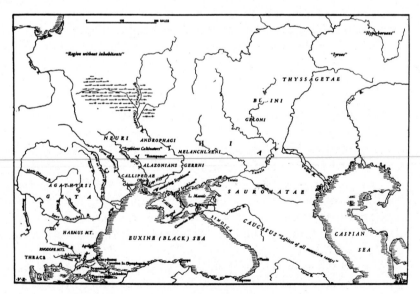

Figure 3. River systems of Scythia according to Herodotus.

propriety, be explained to mean what we now understand by mechanical or chemical attraction, or pressure, or evaporation'.[35] Was Whewell correct? We can only conjecture, but who really knows the truth?

Herodotus also had an interest in various other river characteristics. He described in great detail[36] the river systems of Scythia (figure 3). He also noted that when the Nile was about to rise, the hollows and marshy spots near the river became flooded because of the water percolating through the riverbanks.[37] Possibly the historian's worst errors were in the descriptions he gave of the general courses of the Danube and the Nile. He believed that like the Danube, which flows across Europe from the west toward the east, the upper Nile flowed in that direction also.[38] He confused the great river 'containing crocodiles' with the Niger. Later both Juba II, the King of Mauretania, and Pliny made similar mistakes.[39] Considering that these erroneous ideas continued in one form or another for nearly the next 2200 years, this early historian deserves to be excused.[10]

HIPPOCRATES' CONCEPT OF WATER

It is difficult to believe that the concepts of the Ionian philosophers had for their bases any observed facts or even a limited amount of experimentation. With characteristic aplomb they proclaimed that the ultimate ingredients of all material things in the universe consisted of water, or air, or atoms, or the four elements – as if the whole process of evolution* had taken place before their very eyes. Plato later described those concepts as being no more than a 'plausible tale'. However, things were different with respect to medical science. Hippocrates (460-400? B.C.), the father of medicine, had some definite ideas about the constitution of water.[40] He thought water was comprised of two parts: one part was thin, light, and clear – the other was thick, turbid, and dark coloured. The sun attracted and raised only the lightest and the thinnest part of water – as should be obvious from the salty part which was always left behind. Water could be withdrawn from all things which contained humidity – and

---

* The evolution process, of course, was not known at that time.

there was humidity in everything. Of all waters, he declared rain water was the lightest, sweetest, thinnest, and clearest.

'When attracted and raised up, being carried out and mixed with the air, whatever part of it is turbid and darkish is separated and removed from the other and becomes cloud and mist, but the most attenuated and lightest part is left, and becomes sweet, being heated and concocted by the sun, for all other things when concocted become sweet. While dissipated then and not in a state of consistence it is carried aloft. But when collected and condensed by contrary winds, it falls down wherever it happens to be most condensed. For this is likely to happen when the clouds being carried along and moving with a wind which does not allow them to rest, suddenly encounters another wind and other clouds from the opposite direction: there it is first condensed, and what is behind is carried up to the spot, and thus it thickens, blackens, and is conglomerated, and by its weight it falls down and becomes rain.'[10]

Hippocrates conducted an experiment to show that some portions of water (the thinnest and lightest portions) could be eliminated by evaporation. A measured quantity of water was poured into a vessel and was then exposed to the open air in winter till it became frozen. The next day it was brought into a 'warm situation' until the ice melted, whereupon it was weighed and found to be much less than the original quantity. From this he concluded that he had 'a proof that the lightest and thinnest part became dissipated and dried up by the congelation, but not the heaviest and thickest, for that would be impossible'.[40] A few hundred years ago, Anaximenes had declared his concept on the effect of reduction of temperature on the density, i.e., the hotter, the thinner; the colder, the denser.[41] Had he tried a simple experiment like Hippocrates, he probably would have thought twice before propounding such a general and universal concept. Water when heated becomes vapour and expands; but what happens when it is frozen? Does it contract into a smaller volume as anticipated by this theory? Had he kept a jar of water outside on a wintry night, he would have noticed that instead of contracting it expanded, and possibly would have even split the container. In comparison, the simple experiment conducted by Hippocrates was a major development. It undoubtedly was a step forward in the right direction – that of conceiving methods for conducting scientific investigations.

ARISTOPHANES

The Athenian playright Aristophanes (445?–385? B.C.) ridiculed the then prevailing concept that rain was sent by the almighty god Zeus. The dialogue between Strepsiades and Socrates, as contained in his play *The clouds*, is worth quoting:

'Strepsiades: No Zeus up aloft in the sky!
        Then, you first must explain, who it is sends the rain;
        Or I really must think you are wrong,
Socrates: Well then, be it known, these send it alone:
        I can prove it by arguments strong.
        Was there ever a shower seen to fall in an hour
        when the sky was all cloudless and blue?
        Yet on a fine day, when the clouds are away,
        he might send one according to you.
Strepsiades: Well, it must be confessed, that chimes in with the rest:
        your words I am forced to believe.
        Yet before, I had dreamed that the rain-water streamed
        from Zeus and his chamber-pot sieve.'[42]

CONCLUSION

The tradition of free inquiry started with the Milesian school, notably from the time of the 'first philosopher' Thales, and every physical phenomenon was made the subject of discussion and criticism. It was during this time of the Hellenic Civilization that man first seriously attempted to understand nature, and began giving thought to natural causes rather than divine ones. Thales was the first man to assign much importance to water. His thoughts were echoed later by Pinder, who in the fifth century B.C., flatly stated that the best of all things is water. Here, for the first time in history, man pursued knowledge for its own sake. During this period also, the seeds of hydrology as a science were being sown. They finally blossomed some 2200 years later with the seventeenth century experimental works of Pierre Perrault, Edmé Mariotte and Edmond Halley.

REFERENCES

1. REYMOND, A., Science in Greco-Roman antiquity, translated by R. Ghèury de Bray. London, Methuen & Co. (1927).
2. RAPIN, R., Reflections upon physicks. In: The whole critical works of Monsieur Rapin, translated by Basil Kennet, vol. 2, 2nd ed. London (1716) pp. 449–451.
3. ARISTOTLE, Politics. Book 1, 1259A.
4. ARISTOTLE, De caelo (On the heavens). Book 11, 294a.
5. ARISTOTLE, Metaphysica (Metaphysics). Book 1, chapter 3, 383b.
6. GUTHRIE, W. K. C., A history of Greek philosophy, vol. 1. Cambridge, University Press (1962).
7. FARRINGTON, B., Greek science. Harmondsworth, Penguin Books (1963) p. 37.
8. BURNET, J., Early Greek philosophy. London, Adam and Charles Black (1892) pp. 42–45.
9. OSWIECIMSKI, S., Thales – the ancient ideal of scientist. Warsaw, Charisteria Thaddaeo Sinko (1951) pp. 230–250.
10. BISWAS, ASIT K., The Nile: its origin and rise. Water and Sewage Works *113* (1966) 282–292.
11. ROSE, V., Aristoteles pseudepigraphus. Lipsiae (1893).
12. DIELS, H., Doxographi Graeci. Collegit, recensuit, prolegomenis indicibusque instruxit Hermannus Diels. Berolini (1879) pp. 226–229.
13. SARTON, G., Introduction to the history of science. From Homer to Omar Khayyam, vol. 1. Baltimore, William and Wilkins Co (1927) pp. 206–207.
14. SARTON, G., A history of science. Ancient science through the Golden Age of Greece. Cambridge, Mass., Harvard University Press (1953) p. 559.
15. VITRUVIUS, The architecture of Marcus Vitruvius Pollio in 10 books, translated by J. Gwilt. London, Priestley and Weale (1826) pp. 227–228.
16. JONES, T. B., Ancient civilization, 2nd revised printing. Chicago, Rand McNally & Co. (1966) pp. 265–266.
17. SARTON, G., Ref. 14 pp. 175–177.
18. HEIDEL, W. A., Anaximander's book. The earliest known geographical treatise. Proceedings of the American Academy of Arts, Sciences, and Letters *56* (1921) 275.
19. ARISTOTLE, Meteorologica. Book 2, chapter 1, 353b.
20. DIELS, H., *op cit.*, p. 560.
21. DIELS, H., *op. cit.*, p. 370.
22. FREEMAN, K., Ancilla to pre-Socratic philosophers. (Translation of H. Diels' Die Fragmente der Vorsokratiker, 5th ed.) Oxford, Basil Blackwell (1948) p. 23.
23. GOMPERZ, H., Problems and methods of early Greek science. Journal of the History of Ideas *4* (1943) 161–176.
24. PEASE, A. S., Fossil fishes again. Isis *33* (1942) 689–690.
25. BURNET, J., *op. cit.*, pp. 295–297.
26. FREEMAN, K., *op. cit.*, p. 71.

27. DE SELINCOURT, AUBREY, The world of Herodotus. London, Secker and Warburg (1962) pp. 221–223.

28. HERODOTUS, The history. Book 2, 5.

29. VON ZITTEL, C. A., History of geology and palaeontology to the end of the nineteenth century, translated by M. M. O. Gordon. London, Charles Scribner's Sons (1901).

30. HERODOTUS, *op. cit.*, Book 2, 11.

31. HARRINGTON, J. W., The first principles of geology. American Journal of Science *265* (1967) 449–461.

32. HERODOTUS, *op. cit.*, Book 2, 13–14.

33. HERODOTUS, *op. cit.*, Book 4, 50.

34. BISWAS, ASIT K., Experiments on atmospheric evaporation till the end of the eighteenth century. Technology and Culture *10* (1969) 49–58.

35. WHEWELL, W., History of inductive science, vol. 1. London, John W. Parker & Son (1857) pp. 26–27.

36. HERODOTUS, *op. cit.*, Book 4, 46–58.

37. HERODOTUS, *op. cit.*, Book 2, 93.

38. HERODOTUS, *op. cit.*, Book 2, 32–34.

39. SARTON, G., Introduction to the history of science, vol. 3, part 2. Baltimore, Williams & Wilkins (1947) p. 1158.

40. HIPPOCRATES, Hippocratic writings, translated by Francis Adams. Great Books of the Western World, vol. 10. Chicago, Encyclopaedia Britannica Inc. (1952) p. 12.

41. CORNFORD, F. M., Was the Ionian philosophy scientific? Journal of Hellenic Studies *62* (1942) 1–7.

42. ARISTOPHANES, The clouds, translated by B. B. Rogers. Great Books of the Western World, vol. 5. Chicago, Encyclopaedia Britannica Inc. (1952) p. 492.

3

The age of Plato and Aristotle

INTRODUCTION

The Sophists and the Socratic school dominated the trend of the scientific thought from about the middle of the fifth century to the early fourth century B.C. Our knowledge of this period comes, in general, from the writings of Plato. The Sophists not only thought that the teaching of pure science was degrading but they even attacked it vehemently at times. This condition changed with the advent of Plato who, far from supporting the self-styled 'masters of wisdom' philosophers, professed a great love for mathematics. He was interested in scientific principles and methods – particularly in the elemental structure of the universe. If he had his way, 'there ought to be a law' which would make the study of mathematics compulsory to all would-be statesmen. Inscribed across the top of his academy was the statement: 'Let no man ignorant of geometry enter here'.[1]

No man had ever influenced the development of scientific thought for such a long time as had Aristotle, and it is highly unlikely that his record will ever be exceeded. His opinions had withstood the test of time for nearly two millenniums. Admittedly it had some ups and downs during this period but it is still a remarkable achievement. Aristotle was an encyclopedist, and with the probable exception of Democritus, he was the first encyclopedist of the human race. Much of our scanty knowledge of the Greek philosophy is handed down to us from his writings. He was a disciple of Plato but he far

Figure 1. Plato (by courtesy of Ny Carlsberg Glyptotek, Copenhagen).

surpassed his master in every branch of knowledge. This was true
to such an extent, that some claim Plato became well known largely
because he was the teacher of Aristotle. Plato's contribution to
knowledge was in the form of imaginary conversations between
persons. It is often impossible to distinguish historical facts from

pure fiction, therein. Aristotle's approach was entirely different. He almost invariably assembled the opinions of learned men of bygone days and then added his own contributions to them.

In this chapter Plato's and Aristotle's contributions to the development of the science of hydrology will be briefly examined.

PLATO

His life

According to Alexandrian scholars, Plato (figure 1) was born in the month of *Thargelion* (May–June) of the first year of the eighty-eighth Olympiad (428–427 B.C.). Son of distinguished Athenian parents of the Periclean era, Ariston and Pericrione, he devoted most of his life to the search for truth. It is highly probable that even as a boy, he knew Socrates and his early ambition appears to have been political. The execution of Socrates in 399 B.C. on a trumped up charge of impiety affected him profoundly, and it probably made him give up his political aspirations. He visited Italy and Sicily, and in about 387 B.C., on his return to Athens, he founded the Academy, an event which was a milestone in the history of science and philosophy.

He was invited to educate the new 30-year old King Dionysius II of Syracuse in 367 B.C. Finding that he was not very successful at that task, he returned after a few months to Athens. King Dionysius again persuaded Plato to continue that effort in 361 B.C., but in this second attempt he again failed to induce the King to follow a course of combining both power and philosophy. Within about a year he returned to Athens (at a considerable personal danger) and from then on he never again ventured into politics. He spent the remaining years of his life at the Academy and died in 348 or 347 B.C.

Water – a primary element

Plato accepted the concept of the four basic elements of matter, fire, air, water, and earth, as first postulated by the Greek philosopher Empedocles of Agrigentum, but added thereto a fifth element which he described in somewhat mysterious terms.[2] Aristotle later explained that the fifth element must have been heaven. Proclus in his commentaries quoted Plato's *Timaeus* as saying that 'He who constituted the world composed it from all fire, air and earth, leaving no part nor power of any one of them externally'. Then he adds:

'He does not say from fire or water simply, but from all fire and all water, through which he indicates that there is much fire in the universe, and of a different nature, and also much water, and which is essentially different. More-over, the theology of the Assyrians which was unfolded into light from divinity, delivers the same things. For in that theology, the Demiurgus is said to have made the whole world from fire, water, earth, and all-nourishing ether or air; and the artificier is said to have fashioned the world as it were with his own hands.'[3]

The four *archē* theory accordingly does not seem to have been new – Empedocles apparently borrowed it from the Assyrians.

Plato's main scientific work is considered to be *Timaeus*, which, according to Sarton, has been considered by many commentators as the ultimate of Platonic wisdom which prevailed for the thousands of years thereafter, but which 'modern men of science can only regard as a monument of unwisdom and recklessness'.[4] *Timaeus* is of some importance to hydrologists because it attempts to describe a mathematical theory regarding water – one of the five basic elements of the universe. His thought was that 'as the world must be solid, and solid bodies are always compacted not by one means but by two, God placed water and air in the mean between fire and earth and made them to have same proportion so far as was possible (as fire is to air so is air to water, and as air is to water so is water to earth) ... out of such elements which are in number four, the body of the world was created, and it was harmonized by proportion, and therefore has the spirit of friendship.'[5]

The Athenian mathematician Theaetetos was the first to formulate the theory of polyhedra – often known as Platonic bodies. Since he is the principal character in one of the most famous of Plato's dialogues it is not unreasonable to believe that Plato obtained his concept of five regular solid figures from Theaetetos. The theory of polyhedra stated that there could only be five regular solid figures having as their sides regular identical polygons and without any re-entrant angles. Plato assigned four of the five polyhedra to the four Empedoclean elements. The remaining fifth compound figure was referred to in the following somewhat mysterious terms: 'There was yet a fifth combination which God used in the delineation of

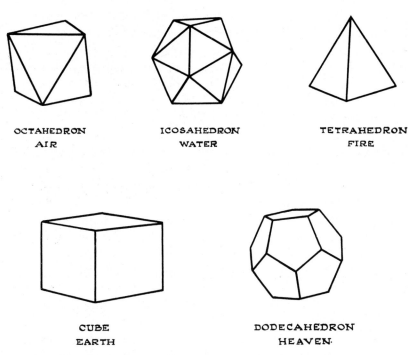

OCTAHEDRON   ICOSAHEDRON   TETRAHEDRON
AIR   WATER   FIRE

CUBE   DODECAHEDRON
EARTH   HEAVEN·

Figure 2. Platonian concept of elements.

the universe'.[6] Thus air consisted of octahedra, water of icosahedra, fire of tetrahedra, earth of cubes, and heaven (?) of dodecahedra (figure 2).[7] The reason for assigning icosahedra to water was because the form was least mobile and had the greatest body.

The Polish logician Wincenty Lutowslawski made a remarkable comment in his book on the *Origin and growth of Plato's logic*. He believed that from Plato's writings one can conclude that[8] water consists of three atoms: two of one gas and one of another – which undoubtedly is stretching the cult of Plato too far. Sarton comments: 'Lutowslawski reminds me of the people who read scientific anticipations of the Bible or the Qur'an'.[4]

Origin of rivers and springs
Two possible explanations are available in the dialogues of Plato on the origin of rivers and springs, of which the most quoted hypothesis[9] is the Homeric ocean (Tartarus) concept. Plato believed

that there are numerous interconnected perforations and passages, broad and narrow, in the interior of the earth.[10] He imagined the existence of a huge subterranean reservoir called Tartarus. This was the vastest of all chasms, and it penetrated the entire earth. The watery element therein had neither a bed nor a bottom; it always surged to and fro. 'When' Plato wrote, 'the water retires with a rush into the inner parts of the earth, as they are called, it flows through the earth into those regions, and fills them up like water raised by a pump. When it leaves those regions, and rushes back hither, it again flows into the nearby hollows, and when these are filled, it flows through subterranean channels and finds its way to several places, forming seas, and lakes, and rivers, and springs'.[10] All waters of rivers and streams flow back to Tartarus directly or through a circuitous route. Like the Egyptians, Plato was aware of one of the fundamental principles of water, that it always flows downhill. Perhaps that was why he stated that the entrance of the rivers back to the earth was always lower than the level at which they originated. The flow from Tartarus to the rivers and vice versa was a continuous process.

The alternate possible explanation of the origin of springs and rivers can be found in *Critias*. Speaking of conditions at Athens some 9000 years before his time, Plato said:

'Furthermore it [the land of Attica in ancient times] enjoyed the fructifying rainfall sent year by year from Zeus; and this was not lost to it by flowing off into the sea, as nowadays because of the denuded nature of the land. The land [then] had great depth of soil and gathered the water into itself and stored it up in the soil we now use for pottery clay, as though it were a sort of natural water-jar; it drew down into the natural hollow the water which it had absorbed from the high ground and so afforded in all districts of the country liberal sources of springs and rivers; and surviving evidence of the truth of this statement is afforded by the still extant shrines, built in spaces where springs did formerly exist.'[11]

The Tartarus concept gives an account of the origin of streams and rivers, and a question arises as to whether it represents Plato's own view or someone else's. Krynine, for example, regarded the Platonic idea of a huge subterranean reservoir as 'an ironical one' on the basis of a statement of Socrates to one of the principal characters of the dialogue *Phaedo*, that 'I can tell you a charming tale, Simmias, which is well worth hearing'.[12] He went on to credit Plato with describing the first pluvial concept of the hydrologic cycle. Krynine then stated that the 'origin of springs and rivers [as] postulated in the *Critias* is based on acute observation and not on a tongue-in-check embellishment of Homeric myths as in the *Phaedo*'.[12] It is true that Plato described the other world (including Tartarus) in its relationship with the immortality of the soul. 'He does this without pretending that his account is exact, but this he says, or something like this, must be true. His picture of the other world is borrowed partly from the mythological tales of the poets and the priests, partly from the physical speculations of the philosophers, and is in a good measure, as we can perceive, expanded and adorned by Plato's own imagination'.[13] Thus, his account was partly mythological and partly literal, and it is almost impossible to distinguish one from the other as the transition has been entirely disregarded. Perhaps Taylor has the last word: 'It is useless to discuss the question how much in these myths of the unseen represents a genuine extra belief of either Socrates or Plato, and how much is conscious 'symbolism'. Probably neither philosopher could have answered the question himself'.[14]

It should be pointed out however, that Aristotle takes Plato to task[15] because of his acceptance of the Tartarus concept of the origin of springs and rivers, without ever having mentioned the pluvial concept as having been previously described in *Critias*. It is difficult to believe that Aristotle could have made such a critical statement unless he was convinced of Plato's belief in the Tartarus idea,[16] despite of the fact that the Stagirite was a very unsympathetic critic of his master.

The passage quoted by Krynine in support of his claim that Plato was the originator of the pluvial concept, is from the Bury translation,[17] but it is extremely difficult to draw the same conclusion from other major translations like the Jowett[18] or the Taylor[19] (the one used in this chapter is a literal translation). The passage concerned is extremely involved, and there are reasons to believe that some of it is corrupt.[20]

Krynine discards the Tartarus idea as parageological, Dantesque, and a 'charming tale', but such a criticism is not valid. In the beginning of his speech Critias invoked the aid of all the Gods, especially Mnemosyne (the Goddess of Memory), because he had heard the 'tale' he was about to narrate from Solon of Athens (635–558 B.C.), one of the seven wise men, who in turn had heard it from an exceedingly old Egyptian priest. Thus if Tartarus is to be discarded as a 'charming tale', then the 'tale' of the pluvial concept should also be discarded on the basis of the same reasoning (assuming that Plato did in fact propose such as hypothesis). Besides the whole story 'has so much the appearance of a myth, that it seems useless to apply to it any of the laws of historical or geographical criticism'.[20] Rivaud believed that much of the denudation of Attica, ascribed by Plato to the natural cataclysm, was actually the work of man.[21] He also pointed out that the tale of the lost island of Atlantis came from Egyptian priests, and as they took great pleasure in deceiving the Greeks, that tale is not surprising. But nowhere does he suspect that 'there is a greater deceiver or a magician than the Egyptian priests, that is to say, Plato himself, from the dominion of whose genius the critic and natural philosophers of modern times are not wholly emancipated'.[22]

Finally, 'Atlantis is a creation of Plato's own imagination – a creation which he knows how to give versimilitude to by connection with the accepted 'scientific' doctrine of terrestrial catastrophies'.[23] Thus, the reader of the *Critias* must bear in mind that 'the geology of that work is, after all, the geology of the Aetiological Myth, in which a result, which Plato, as scientific observer, may well have conceived as due to a secular process, was bound to be attributed to a 'catastrophie'.[23] Sarton comments: 'Many geologists have wasted their time in trying to give some appearance of reality to Plato's dream',[22] but whatever may be the merits or the demerits of the dialogue *Critias*, it does indicate that Plato did take some interest in the observation of natural processes.[24]

Thus, it is concluded that it is futile to disregard the Tartarus theory of Plato and credit him with only the pluvial concept of the origin of springs and rivers, as suggested by Krynine. He should be credited with both; assuming that he did put forward a second hypothesis.

Water laws

Plato was interested in water laws and his ideas will be briefly discussed herein. In general, he was satisfied with the old laws, and did not see any good reason for changing them.

Anyone was permitted to draw water from a common stream on his land as long as he did not cut off the flow of a private stream. The water could be caused to flow in any direction except through a house, temple, or sepulchre, but one needed to be careful not to do any harm in excess to the excavation of the channel. In case of water deficiency, one should dig down to the clay layer, and if still no water was found, he had the right to obtain water for his household from his neighbour. If his neighbour's supply was limited, he was permitted to obtain from him an amount as determined by a warden.[25] A man living on higher ground was not permitted to allow the runoff resulting from a heavy rain to drain recklessly on to the land of his lower neighbour, nor could the lower neighbour refuse to furnish an outlet for reasonable drainage from the higher land. In case of dispute the warden would decide what would be required of each man. If any one intentionally polluted or wasted the water of a stream or reservoir of another by poisoning, digging, or by theft, he would be required to pay damages equal to the value of the loss. If he had polluted the water, he also had to purify the water. If the previous Platonic concept had been followed continuously throughout the ages there would have been no water pollution problems for us to contend with at present!

ARISTOTLE

His life

Aristotle (figure 3) was born in 385 B.C. in the city of Stagira in Macedonia. His father, Nicomachus, was a celebrated physician at the court of Amyntas III[26] (the father of Philip of Macedon). He entered Plato's Academy in 367 B.C., and left (some 20 years later) after the death of his teacher. At the invitation of King Philip, Aristotle became tutor to the young prince Alexander. About seven years were devoted to teaching Alexander politics and rhetoric. After the accession of Alexander, Aristotle returned to Athens, where, because Xenocrates denied him the headship of the Academy, he founded the Lycaeum, later renowned as the Peripatetic School, which was an immediate success. Like Plato, he was inter-

Figure 3. Aristotle (by courtesy of National Museum, Naples).

ested in knowledge for its own sake. After the death of Alexander a revolt took place in Athens, and he was threatened with false prosecution. Remembering the fate of another great philosopher, Socrates, he promptly fled to Chalcis and died there after a few months in 322 B.C.

Concept of water

The Aristotelian concept of the universe is somewhat similar to that of Pythagoras and Plato. He believed in five elements and each element was credited with two of the following four qualities: cold, hot, dry, and humid. Four 'earthly' elements thus had four qualities cold-dry (earth), cold-humid (water), hot-dry (fire), and hot-humid (air). They could engender each other in a circular way (figure 4), and he quoted Empedocles in support of his theory:

'The sun everywhere bright to see, and hot;
The rain everywhere dark and cold.'[27]

Aristotle's universe was unique and finite. Each element had its appropriate place in the universe, and its nature was to move toward

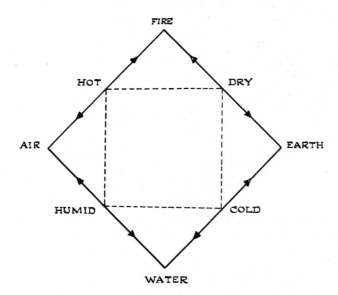

Figure 4. Aristotelian concept of elements.

such a place. Having reached there, it would cease to move any further. He believed the earth was at the centre of the universe, and that it was surrounded by water, whereas fire was below heaven but above air. Nothing existed beyond the highest heaven – not even a surrounding void, because space was finite.[28] Figure 5 is a medieval representation of the elements of the earth according to Aristotle.[28]

Figure 5. Position of elements according to Aristotle (from Fine, *La théorie des ciels, mouvemens, et termes pratiques des sept planètes*, 1528).

First book on meteorology
Aristotle was the first man to write a treatise on meteorology. It was entitled *Meteorologica*. Not only did it deal with meteorology,

using the root of the modern word, but also with astronomy, chemistry, geology, and physics. It is quite likely that the fourth book of *Meteorologica* was written by Straton,[29] and it may be considered to be the first elementary text on chemistry to have been written. However, the portions of the book most likely to interest the hydrologists deal with rain, dew, hail, snow, wind, ocean, and the origin of rivers and springs.

Mechanics of precipitation

Aristotle, like his predecessors, did not have a clear concept of the difference between air and vapour. Air, he said, being hot and moist, is a sort of aqueous vapour. He believed that two types of evaporation took place from the earth's surface due to the sun's heat: one was evaporation, and the other was a kind of windy exhalation which was like smoke, but which arose from the dry earth. The windy exhalations being drier and warmer, would rise above the moist vapour. Aristotle argued that as the sun approached or receded from a place, it gave rise to the dissipation or condensation of vapour, which were forms of degeneration and destruction. He believed that the heat responsible for evaporation came from the sun and also from the 'other heat from above',[29] but he did not explain his second source.

The water vapour gradually loses heat as it rises, partly due to its rising far into the upper air, and partly due to dispersion into the higher region, with the result that it condenses and changes from 'air into water'.[30] The condensed 'air' thus forms clouds which in turn produces rain. Mist is formed from what is left over after the cloud has been turned into water. In other words, mist is a form of 'barren cloud', and hence is a symbol of good weather rather than of rainfall. As the sun approaches a particular locality, evaporation takes place, and as it recedes, condensation and rainfall occur – thus forming a cyclical process which follows the course of the sun. Then followed what was no doubt Aristotle's conception of what we call the hydrologic cycle:

'Some of the vapour that is formed by day does not rise high because the ratio of the fire that is raising it to the water that is being raised is small. When this cools and descends at night, it is called dew and hoar-frost. When the vapour is frozen before it has condensed to water again it is hoar-frost ... It is dew when the vapour has condensed into water and the heat is not so great as to dry up the moisture that has been raised, nor the cold sufficient (owing to the warmth of the climate or season) for the vapour itself to freeze.'[31]

Rainfall results from the condensation of a great amount of vapour from a considerable area over a long period of time, whereas dew is produced from the vapour collected in a single day from a small area as indicated by its quick formation and scanty quantity. The freezing of cloud produces snow. Hoar-frost occurs as the product of frozen vapour. The relationships between rain and dew, and snow and hoar-frost are accordingly similar; that is, there is a great deal needed to produce the former, and very little to produce the latter.

Aristotle believed that the intensity of rainfall was directly dependent on the rate of condensation of water vapour, and that heavy showers would occur as a cloud descended into warm air. This was exactly opposite from the view expressed by Anaxagoras.

Origin of springs and rivers

Aristotle was critical of Anaxagoras' concept of the origin of rivers. Anaxagoras had envisaged a subterranean reservoir or reservoirs from which all rivers originated, and that such rivers were perennial if the reservoirs in question could store enough water to last until the arrival of the winter rains. Without adequate storage, the rivers just dried up for some periods during the year. Aristotle, to the contrary, contended that if a single reservoir existed which was capable of storing all the water flowing in the rivers continuously day by day, its size would have to be larger than the earth itself, or under most circumstances, not much smaller.[32]

Having thus disposed the Anaxagorean theory, Aristotle derided the concept of Tartarus as propounded by his teacher Plato. He argued first, that if that theory were correct, rivers would flow in a direction which conformed with the surging of Tartarus, and that this direction was not necessarily downwards in all instances. The proverbial case of a river flowing upward could then become a reality; second that the Tartarus theory does not take into account generation of new water – that if all the water that flowing out of Tartarus goes back there again, how can the loss of water due to vapourization be accounted for? The treatise gives the impression that Aristotle was very surprised to find that one could be stupid enough to make the mistake of neglecting the formation of new water. His final argument was that none of the rivers on or in the earth go back to Tartarus but that they in fact discharge either into the sea or into one another. He thereupon concluded that all these factors make the Platonic concept of a huge subterranean reservoir clearly impossible, particularly so, because the theory about the subterranean sea was supposed to have originated from Tartarus. Having thus discarded Anaxagorean and Platonic postulates, Aristotle went on to submit his own views on the subject. He said that if cold changes air into water above the earth, it must also produce the same effect within the earth. Thus there is continuous conversion of air into water inside of the earth, and it would be extremely unreasonable for anyone to refuse to admit it.

'Just as above the earth, small drops form and these join others, till finally water descends in a body as rain, so too we must suppose that in the earth the water at first trickles together little by little and that the sources of rivers drip, as it were, out of the earth and then unite. This is proved by facts. When men construct an aqueduct they collect water in pipes and trenches, as if the earth in the higher ground was sweating the water out. Hence, too, the headwaters of rivers are found to flow from mountains, and from the greatest mountains there flow the most numerous and greatest rivers. Again, most springs are in the neighbourhood of mountains and of high ground, whereas if we except rivers, water rarely appears in the plains. For mountains and high ground, suspended over the country like a saturated sponge, make the water ooze out and trickle together in minute quantities but in many places. They also receive a great deal of water falling as rain.'[33]

Another explanation for the origin of rivers and springs is given by Aristotle in the continuation of the passage quoted above, viz.: 'they *also* cool the vapour that *rises* [the author's italics] and condense it back into water'.[34] The passage is rather obscure and, according to Kircher,[35] there was considerable controversy in the middles ages about the origin of the vapour that rises, as nowhere in his treatise did Aristotle mention where it came from.

Thus Aristotle had three different explanations for the origin of rivers:

(1) rainfall and percolation;
(2) subterranean condensation of 'air' into water; and
(3) condensation of vapours rising (from some source not stated).

He held that the mountains act as spongy receptacles. If these reasons, single or in combination, truly explain the origin, what would make the river perennial or seasonal? Aristotle was ready with his answer. In brief, the reason which he gave lies in the size of the mountains and their density and cold temperatures. He contended that if mountains are big, dense, and cold, they can catch, create, and retain most of that water, and hence the rivers which originate within them will be perennial. If on the contrary, they are small, or porous and stony and clayey, they will soon lose their supply of water, and the resulting rivers will consequently have only seasonal flow.[36]

Saltness of the sea
Aristotle, like Hippocrates, believed that only the lighter and the sweeter portions of water were evaporated by the sun. He argued that the sea must persist forever because as the sun approaches water, it is drawn up, and as the sun recedes, the water comes down as rain. As long as this arrangement continues the sea should never dry up.[37]
When the water of the sea evaporates, it somehow rids itself of its salty taste. He could vouch for that from the experiments he conducted.[38] This phenomenon, he declared, is true of all wines and fluids. When distilled, they become water.[39] His explanation for the salinity of sea-water was strange indeed, however, because he believed that such salinity was due to the earthy residue always left by everything that grows and is naturally generated in this world. Since moist and dry evaporations are mixed, some quantity of it must exist in the clouds, and hence when precipitation occurs, it carries with it the saline dry exhalation. This reason was also given for the salinity of the first autumnal rains and rains coming from the south. As this process is repeated continuously the sea gradually became salty.

CONCLUSION

Plato's major contribution to hydrology was the two hypotheses he put forward for the origin of rivers and springs. Of the two, the Tartarus concept has been the most quoted one, and Plato was severely critized for it by Aristotle. It appears that Plato had also put forward the pluvial concept in his work *Critias* which has received very little attention so far. It is suggested that Plato should be credited with both the Tartarus and pluvial hypotheses on the origin of rivers and springs.

It is extremely doubtful if any one could successfully argue against the fact that Aristotle was one of the greatest intellects of all time.[40] His interests had an encyclopedic range, and he had grasped the principles of scientific method more than any other man before and long after his time. His approach to knowledge was rather pedantic, and his greatest contributions were in the field of biology. The Aristotelian School collected a large number of observations from which they derived many intelligent interpretations, but often they were wordy and inconclusive. Sometimes he finished his explanations with the words *kai para tauta uden* (and beyond that nothing else). He did pass along some old-wive's tales such as men have more teeth than women, and although he had been twice married, it apparently never occurred to him to check the truth of that tale by making observations. It may be said that the Greek did not have a knowledge of measuring instruments, and hence should be excused for resorting to speculations, but this condonation would be all too generous.[41] Undoubtedly they were capable of measuring length (linear straightedge), time (water-clock), and weight (balance), but they did not make any general use of them. They were more interested in 'why?' than 'how much?' Aristotle stated that 'All men by nature desire to know' but mankind had to wait for nearly another two millenniums for Johann Kepler to add 'to *measure* is to know'.

Perhaps the contributions of Plato and Aristotle to the science of hydrology can be best summed up by the words of Aristotle which are engraved in the National Academy of Sciences Building in Washington, D.C.:

'Search for truth is one way hard and in other way easy, for it is evident that no one can master it fully nor miss it wholly, but each adds a little to our knowledge of nature, and from all the facts assembled there arises a certain grandeur.'

REFERENCES

1. BISWAS, ASIT K., In defence of Plato. Engineering *200* (1965) 391.
2. MICHEL, P. H., The Sophists, Socrates and Plato, part II, ch. 3. In: Ancient and medieval science, edited by R. Taton, translated by A. J. Pomerans. London, Thames and Hudson (1964) p. 225.
3. PROCLUS, The commentaries of Proclus on the Timaeus of Plato, translated by Thomas Taylor, vol. 1. London (1820) p. 427.
4. SARTON, G., A history of science, ancient science through the Golden Age of Greece. Cambridge, Harvard University Press (1959) pp. 420–423.
5. PLATO, Timaeus, translated by B. Jowett, vol. 3, 3rd ed. Oxford, University Press (1892) p. 353.
6. PLATO, Timaeus. In: The dialogues of Plato and the seventh letter, translated by B. Jowett. Chicago, Encyclopaedia Britannica Inc. (1952) p. 459.
7. BISWAS, ASIT K., Atmospheric evaporation: development of concept and measurement up to the end of the 18th century. Technology and Culture *10* (1969) 49–55.
8. LUTOWSLAWSKI, W., The origin and growth of Plato's logic, with an account of Plato's style and of the chronology of his writings. London, Longmans and Co. (1893) p. 484.
9. BISWAS, ASIT K., The hydrologic cycle. Civil Engineering, ASCE *35* (1965) 70–74.
10. PLATO, Phaedo. Dialogues of Plato, translated by B. Jowett, vol 2, 3rd ed. Oxford, University Press (1892) pp. 259–266.
11. PLATO, Critias, 111d.
12. KRYNINE, P. D., On the antiquity of 'sedimentation' and hydrology. Bulletin of the Geological Society of America *71* (1961) 1712–1725.
13. WHEWELL, W., The Platonic dialogues, vol. 1, 2nd ed. London, Macmillan and Co. (1860) p. 419.
14. TAYLOR, A. E., Plato: the man and his work, 7th ed. London, Methuen and Co. Ltd. (1960) p. 207.
15. ARISTOTLE, Meteorologica, translated by E. W. Webster. Chicago, Encyclopaedia Britannica Inc. (1952) p. 461.
16. BURNET, J., Greek philosophy: Thales to Plato, part 1, 1st ed. London, Macmillan and Co. Ltd. (1914) pp. 312–313.
17. PLATO, Critias, translated by R. G. Bury. Loeb Classical Library. London, William Heinemann Ltd. (1929) pp. 273–275.
18. PLATO, Critias, translated by B. Jowett, vol. 3, 3rd ed. Oxford, University Press (1892) p. 532.
19. PLATO, The Critias or Atlanticus. The works of Plato, vol. 2, translated by T. Taylor and F. Sydenham, printed for T. Taylor by R. Wilks. London, Chancery-Lane (1804) p. 581
20. PLATO, Critias. The works of Plato, vol. 2, translated by H. Davis. London, G. Bell and Sons (1916) p. 419.

21. RIVAUD, A., Timée, Critias. Platon, oeuvres complètes, vol. 10. Paris, Société d'édition 'Les Belles-lettres' (1925) p. 239.
22. SARTON, G., *op. cit.*, p. 422.
23. STEWART, J. A., The myths of Plato. London, Centaur Press Ltd. (1960) pp. 417–418.
24. PLATT, A., Plato and geology. Journal of Philology *18* (1889) 134–139.
25. PLATO, Laws, 843, 845.
26. HART, B., Makers of science. Oxford, University Press (1924) p. 25.
27. ARISTOTLE, De generatione et corruptione, translated by H. H. Joachim. Great Books of the Western World, vol. 8. Chicago, Encyclopaedia Britannica Inc. (1952) p. 315.
28. FINE, ORONCE, La théorique des cielz, mouvemens et termes pratiques des sept planètes. Paris, J. Pierre (1528).
29 SARTON, G., *op. cit.*, p. 518.
30. ARISTOTLE, Meteorologica, I.9, 346b.
31. *Ibid.*, I.10, 347a.
32. *Ibid.*, I.13, 349b.
33. *Ibid.*, II.2, 356a.
34. *Ibid.*, I.13, 350a.
35. KIRCHER, A., Mundus subterraneus. Amstelodami, Apud J. Janssonium et E. Weyerstraten (1665).
36. ARISTOTLE, Meteorologica, I.14, 352b.
37. *Ibid.*, II.3, 356b.
38. *Ibid.*, II.3, 358b.
39. FOBES, F. H., Textural problems in Aristotle's Meteorology. Classical Philology *10* (1915) 188–214.
40. GUTHRIE, W. K. C., Aristotle as a historian of philosophy: some preliminaries. The Journal of Hellenic Studies *77* (1951) 35–41.
41. BOYER, C. B., Aristotle's physics. Scientific American *182* (1950) 48–51.

4

The post-Aristotelian period

INTRODUCTION

After Aristotle, the headship of the Lyceum was taken over by his pupil and almost contemporary Theophrastus (371/370–288/287 B.C.). According to Diogenes Laertius he was called Tyrtamus in his early days but Aristotle was so impressed with his eloquence that he named him Theophrastus (the divine speaker). He was a voluminous writer and Diogenes ascribed the authority of some 227 treatises to him on all sorts of subjects. Nearly all his works are lost but one can draw some conclusions from what are available at present – fragments of treatises on *De signis tempestatum (pluvarium, ventorum, tempestatis et serenitatis)*, *De ventis*, and *Meteorologica*.

Somewhere around the time of Theophrastus, the first ever quantitative measurements of rainfall were being taken in India. The credit for the measurements goes to a resourceful Chancellor of Exchequer, Kautilya, who decided to tax the land according to the precipitation it received which presumably would be an indirect form of taxation on agricultural products.

THEOPHRASTUS ON HYDROLOGY

The original work of Theophrastus on meteorology is lost. It was, however, translated into Syriac, and was later retranslated into Arabic by an anonymous Arab who also made an abstract of his translation. Bergsträsser[1] believed that the abstract was made in

1446 or 1447 A.D. by an unlearned and careless scribe from an
Epicurean source of Theophrastus' work on meteorology. On the
whole, it is presently accepted that the four pages of the available
abstract are of Theophrastian origin.[2] It indicates that the meteorological concepts of Theophrastus followed the conventional pattern
of that time, and probably begins at the point from which his great
master left off. Some of the statements from his abstract, with brief
discussions, will follow.

Snow falls when the clouds are frozen as they pass through cold
before they have a chance to discharge their load of water. Present
in the cloud are very small drops of water separated by air, and
when freezing takes place snowflakes are formed with the drops containing some air caught in the process. When squeezed and com-
pressed, snow turns into water but its bulk is reduced as the air es-
capes. The lightness and whiteness of the snow further supports the
contention as it is due to the presence of the entrapped air. Frost and
ice are produced due to congelation of dew through cold. Air again
is the cause of whiteness of the ice and it is in fact a combination of
snow on frost. Hail is caused when drops of water are congealed
owing to moisture, and are round in shape either because the corners
are broken off and smoothened as they fall (compare with Seneca's
concept in chapter 5) or else due to the cold which at once compres-
ses it into that form.

It can be said, with some justification, that Theophrastus was the
first man to have a correct understanding of the hydrologic cycle,[3]
and that the Roman architect Vitruvius later restated his ideas. In
the abstract, the reason for the vapour to depart from water surfaces
is attributed to the wind: 'the air often contracts, sometimes in the
east, sometimes in the west, or the north or the south; if it gathers
there and finds no empty space, the air flows from one region to another lying opposite to it, because the empty space attracts it and
with it withdraws vapour from the water and from the earth until
there is no empty space left'.[1] Later Proclus in his commentary on
Timaeus of Plato credited Theophrastus with saying, 'this ... is one
cause of rain, viz. the pressure of clouds against a mountain'.[4] Still
later, exactly the same reasons were expressed in almost the same
words (see chapter 5) by Vitruvius who stated that he too had studied
the works of Theophrastus carefully. It makes one strongly suspect

that the concept of hydrologic cycle was first postulated by that Greek, and Gilbert[5] has made a strong case for such a suspicion. In the beginning of his treatise *De ventis* (On winds), Theophrastus stated categorically that the causes and origin of winds have already been considered, and hence he will only deal with the effect of winds and the forces and conditions associated with them. He was probably referring to his own treatise *Meteorologica* (or conceivably to a work of the same name by his illustrious master). The third book, entitled *De signis tempestatum (pluviarum, ventorum, tempestatis et serenitatis)*, or 'On weather signs (rain, wind, storm, and fair weather)', is a collection of notes concerning weather forecasting. A typical one reads: 'If cranes fly early and in numbers there will be an early storm; but if late and for a long time, the storm will come later. And if they wheel in their flight they indicate a storm'.[6]

HYDROMETEOROLOGICAL OBSERVATIONS

Probably the Greeks were the first people to make systematic hydro-meteorological observations. Theophrastus stated that many men were interested in meteorological observations in Greece as well as in Asia Minor. Such observations were not intended to include quantitative measurements of weather, but rather to present a mere statement of fact such as 'it rained on such and such a date'. At times the hydrometeorological observations were publicly exhibited in the form of *parapegma* (astronomic tables or almanacs). This practice can be traced from the time of Meton and Euctemon in Athens around the fifth century B.C. A typical *parapegma* reads:[7]

September 5 – Rising of Arcturus. South wind, rain and thunder.
September 12 – The weather will likely change.
September 14 – Mostly fine weather for seven days, thereafter easterly winds.

The first quantitative hydrometeorological measurement (that of rainfall) however, dates back to the fourth century B.C. – to the celebrated minister of Chandragupta, founder of the Maurya dynasty of India.

RAIN GAUGE OF KAUTILYA

The earliest reference to a rain gauge was made by Kautilya in

his book entitled *Arthasastra* (the science of politics and admini-
stration), which was probably written towards the end of the fourth
century B.C. It is stated therein that: 'In (front of) the store house,
a bowl with its mouth as wide as an *aratni* shall be set up as a rain
gauge'.[8] He also stated in a later chapter entitled 'The superintendent of agriculture' that:

'The quantity of rain that falls in the country of jangala [forests] is 16 dronas,
half as much more in moist countries; as to the countries which are fit for agriculture – $13^{1}/_{2}$ *dronas* in the country of asmakas [Maharastra]; 23 *dronas* in
avanti; and an immense quantity in Western countries, the borders of the Hima-
layas, and the countries where channels are made use of in agriculture. When
one-third of the requisite quantity of rain fails both during the commencement
and closing months of the rainy season and two-thirds in the middle, then the
rainfall is [considered] very even.'[9]

Aratni was a normal unit of measure comparable to the cubit in
length, about 18 in. – the distance from the elbow to the fingertips.
The unit is still in use in certain parts of rural India but is now
known as *hath* (hand). Unfortunately, Kautilya does not say anything about the shape of the rain gauge. In a comparatively recent
translation of *Arthasastra*, the translator stated in a footnote that
'On the basis of about 511 cubic inches in a *drona* and a cylindrical
rain gauge with a surface area of about 254.3 sq. inches, 16 *dronas*
amount to be about 32 in. of rain; if the gauge-mouth is understood to be square (18 in. × 18 in.) they would amount to be
about 25 in.'[10] According to Webster's dictionary, 1 *drona* is equivalent to 16.5 litres, and assuming 1 litre to be about 61.02 cubic
inches, 1 *drona* would be equivalent to 16.5 × 61.02 = 1007 cubic
inches. Under the circumstances the translator's figures represent
only approximately one-half of the amount computed from the
dictionary's definition.

Kautilya believed that forecasts of rainfall could be made from
observations of the planets Jupiter and Venus, and of the Sun. His
classification of clouds is of considerable interest.[11, 12]

'Three are the clouds that continuously rain for seven days; eighty are they
that pour minute drops; and sixty are they that appear with the sunshine – this
is termed rainfall. Where rain, free from wind and unmingled with sunshine,
falls so as to render three turns of ploughing possible, there the reaping of good
harvest is certain.'[9]

The need for rain gauges arose for two reasons – lands were taxed according to the amount of rainfall[13] they received every year, and the fact that the Superintendent of Agriculture should have, of necessity, a good knowledge of rainfall for planting crops.[14]
The next quantitative measurements of rainfall were made in Palestine around the first century A.D.

CONCLUSION

Two obvious major developments during this period are the development of the concept of the hydrologic cycle, and the first quantitative measurements of rainfall on a rational basis. It is evident that Kautilya maintained a series of rain gauges over the subcontinent of India, and it is a great pity that the system was not continued after his time.

Philo of Byzantium[15] lived sometime after Ctesibius, who flourished around 283–247 B.C.(?), but before Vitruvius. He was aware of the need for repeating experiments[16] when formulating concepts of scientific progress.

'And the ancient did not succeed in determining this magnitude by test, because their trials were not conducted on the basis of many different types of performance. But the engineers who came later, noting the errors of their predecessors and the results of subsequent experiments, reduced the principle of construction to a single basic element . . .'[17]

Philo was an 'ordinary' engineer and a practical man, not a philosopher. It is consequently not surprising that his idea did not get the support it deserved. Had Philo's concept become popular, and had the philosophers accepted his recommended procedure, the history of the development of hydrology or in fact that of any other science would in all probability have been quite different.

REFERENCES

1. LUCRETIUS CARUS, TITUS, De rerum natura, vol. 3, edited by Cyril Bailey. Oxford, Clarendon Press (1947) pp. 1745–1747.
2. STAHL, W. H., Roman science. Madison, University of Wisconsin Press (1962) p. 99.
3. MIDDLETON, W. E. K., A history of the theories of rain. London, Oldbourne Press (1965) p. 10.

4. PROCLUS, DIADOCHUS, Commentary of Proclus on Timaeus of Plato, translated by T. Taylor. London (1820) p. 101.

5. GILBERT, O., Die meteorologischen Theorien des griechischen Altertums. Leipzig, B. G. Teubner (1907) pp. 425–427.

6. THEOPHRASTUS, On winds and on weather signs, translated by J. G. Wood. London, Edward Stanford (1894) p. 67.

7. HELLMANN, G., The dawn of meteorology. Quarterly Journal of the Royal Meteorological Society *34* (1908) 221–232.

8. KAUTILYA, Arthasastra, translated by R. Shamasastry. Bangalore, India, Government Press (1915) p. 64.

9. STAHL, W. H., *op. cit.*, p. 143.

10. KAUTILYA, Arthasastra, translated by R. P. Kangle. Bombay, University of Bombay Press (1963) p. 171.

11. SAMMADAR, J. N., Indian meteorology of the 4th century B.C. Quarterly Journal of the Royal Meteorological Society *37* (1912) 65–66.

12. ANONYMOUS, The antiquity of the raingauges. Scientific American *144* (1912) 236.

13. ANONYMOUS, Irrigation in India through the ages. Central Board of Irrigation of India (1951).

14. BISWAS, ASIT K., Development of raingauges. Journal of Irrigation and Drainage Division, ASCE *93* (1967) 99–124.

15. CLAGETT, M., Greek science in antiquity. London, Abelard – Schuman Ltd. (1957) p 76.

16. EDELSTEIN, L., Recent trends in the interpretation of ancient science. Journal of the History of Ideas *13* (1952) 573–604.

17. COHEN, M. R. and I. E. DRABKIN, A source book in Greek science. Cambridge, Harvard University Press (1948) p. 318.

5

The Roman Civilization

INTRODUCTION

The expression Roman Civilization, as used in the present chapter, refers to that period which extended from about 100 B.C. to the end of the second century A.D. Perhaps it should have been called the Greco-Roman Civilization, because the Romans had relatively very few new independent conceptions to offer even though they managed to build magnificent aqueducts to supply Rome with millions of gallons of water daily, remarkable sewer systems, and a very fine harbour. Even during the peak of the Roman Civilization, the language of learned men was Greek, and all the major writers of the time (like Varro, Vitruvius, Celsus, Pliny, and Seneca) preferred to demonstrate an encyclopedic knowledge rather than to express original and independent thoughts. This had a profound effect on the intellectual life in Western Europe throughout the early Middle Ages.

It can be said, with some justification, that the Romans were 'practical' engineers – for example, their awe-inspiring aqueducts were built without any conscious application of physical principles or unique solutions of constructional problems. Men like Vitruvius and Frontinus did try to lay down some practical principles, but as far as the Romans were concerned, they were satisfied with the existing state of affairs.

VITRUVIUS

His life

Vitruvius was born in northern Italy, possibly at Verona, and except for what can be learned about his life from his famous treatise *De architectura libri decem*, he is practically unknown. His book was dedicated without further identification to the 'imperator Caesar'. Probably the emperor in question was Octavianus, the adopted son of Julius Caesar.[1] It is generally agreed that *De architectura* was written sometime between 27 and 17 B.C.[2] Since he described himself as an old man in his book, it can be inferred that he was at his prime during the second half of the first century B.C. Some have claimed that the book was written around 400 A.D. by a sort of 'pseudo-Vitruvius',[3] who was an ordinary compiler with neither the intelligence to interpret his source nor the literary skill to express himself. It must be pointed out, however, this is a very minority and rather ludicrous opinion.

The book, written in Latin even though the prevalent learned language of the time was Greek, is a diffused compilation on architecture, and Vitruvius was most certainly not a lover of Muses. It dwelt heavily on the Hellenistic tradition rather than the Roman and had a profound effect on the classical architecture, even as late as the Renaissance. Vitruvius had a multi-disciplinary approach to that subject, and advocated that it be combined with knowledge of astronomy, drawing, geometry, history, law, medicine, music, optics, and philosophy. He himself seems to have been a rather unsuccessful architect but he expressed a high regard for himself in the book: 'I promise and expect that in these volumes I shall undoubtedly show myself of very considerable importance not only to builders but also to scholars'.[4] His ambition was undoubtedly more than fulfilled as almost all the earlier theories and practices of renaissance pseudoclassical architecture were based on it. Celebrated architects like Bramante, Michelangelo, and Vignola were considerably influenced by that treatise.

How to find water

Book 8 of his treatise is devoted to water. At the end of its preface he stated that:

Figure 1. Vitruvius' method for locating water.

'As it is the opinion of physiologists, philosophers and priests that all things proceed from water, I thought it necessary as in the preceding seven books [in which] rules are laid down for buildings, to describe in this the method of finding water, [and to mention how] its different properties vary according to the nature of places; how it [the search] ought to be conducted, and in what manner it should be judged, inasmuch as it is of infinite importance for the purposes of life, for pleasure, and for our daily use.'[5]

Chapter 1 is devoted to ways of finding water. When surface springs are not available, water has to be sought and collected from underground sources. The test suggested for locating underground water is to lie flat on the ground before sunrise (figure 1) in the area where water is to be sought, and with one's chin on the ground, to take a close look at the country side, the reason being that the search will then be limited approximately to the same level on the ground. Water can be expected to be found in places where vapours arise from the earth (an impossible condition if the soils were dry). Water may also be sought in localities where there are plants of a type which generally grow in marshy areas. When a promising location is found a hole, not less than three feet square and five feet deep, is to be dug and, a bronze or lead vessel with its inside smeared with oil, is to be placed upside down in that hole at about the time of sunset. The hole is then to be filled with rushes or leaves and earth. If drops of water are found within the vessel on the subsequent day, water should be found at that location. It should be pointed out that nowhere in his book does Vitruvius advocate using a divining rod. In fact, all of his methods for locating underground sources of water have rational bases.

Soil types also provide some idea about the presence of ground water, and hence the nature of ground should be studied carefully. Details regarding the availability of water in various types of soils (according to Vitruvius) are shown in table 1.

A well is to be sunk when a site is located where the experiments mentioned above indicate the possibility of existence of water. Several additional wells are then sunk around the original well and by a series of underground channels water must be brought into a single place from which it can be carried to other locations as needed. Vitruvius suggested that water can be best sought in mountainous regions because water found there will be sweet, whole-

TABLE 1

Details of water available in various types of soils.

Type of soil	Depth at which water may be available	Amount	Taste	Remarks
Clay	Near the surface	Scanty	Not good	—
Loose gravel	Lower down the surface	Scanty	Unpleasant	Muddy
Black earth	—	—	Excellent	Available after winter rains
Gravel	—	Small and uncertain	Unusually sweet	—
Coarse gravel, common sand and red sand	—	More certain	Good	—
Red rock	—	Copious	Good	Difficult to obtain due to percolation
Flinty rock and foot of mountains	—	Copious	Cold and wholesome	—

some, and abundant. There is also no loss of water in such regions due to evaporation as they are 'turned away from the sun's course' and the presence of a forest makes it impossible for the sun's rays to reach the surface water.

The hydrologic cycle
Vitruvius was familiar with the meteorological writings of Aristotle and Theophrastus, and he had a reasonably clear concept of the

hydrologic cycle. He said that valleys between mountains are sub-jected to much rainfall, and snow remains on the ground there much longer because of the dense forests. When the snow melts, it percolates through the interstices of the earth and finally reaches the lowest spurs of the mountains 'from which product the stream flows and bursts forth'.

Vitruvius, like Hippocrates and Aristotle, believed that only the thinnest, the lightest, and the most subtly wholesome part of water was evaporated, and that the heaviest, the harsh, and the un-pleasant parts were left behind. The moisture arising from the earth, during the sunrise, drives the air before it, and in turn receives impetus from the air which rushes in behind it. That onrushing air drives the vapour in every direction – thus creating gales, blasts, and eddies of wind. As the wind travels, it collects further moisture from springs, rivers, marshes, and the sea because of evaporation due to the sun's heat. The 'condensed vapour' then rises to form clouds. Clouds are supported on a 'wave of air', and precipitation occurs when they hit mountains because of the shock sustained and because of their fullness and weight. That explains why there is always more rainfall near mountains than near plains. The rise of vapours, moisture, and clouds from the earth, therefore, is seemingly dependent on the earth's retention of intense heat, great winds, coldness, and on the presence of large amounts of water. 'Thus when, from the coolness of the night, assisted by the darkness, winds arise, and clouds are formed from damp places, the sun, at its rising, striking on the earth with great power, and thereby heating the air, raises vapours and dew at the same time.'[6]

The process is comparable to that of a hot bath where water being heated vapourizes, and the rising vapour forms droplets on the ceiling. When the droplets become large enough they fall on the head of bathers. It is reasonable to assume that since there is no source of water on the ceiling, the water must have come from the bath. The various concepts of Vitruvius were later copied and recopied, and the analogy of the bath house has been repeated as late as the tenth century A.D.

The reasons ascribed by Vitruvius for the origin of hot and cold springs were somewhat analogous to the ones postulated later by Kircher[7] during the seventeenth century. The Roman architect believed that fires are somehow kindled in alum, asphalt, or sulphur in the earth, and that they heat the soil above them. If it so happens that there is a spring in the upper stratum, the water gets heated and produces a hot spring. If the stream travels a long distance after passing through the heated region, the water cools off by the time it reaches the surface, but its taste, smell and colour becomes spoiled in the process.

The final chapter deals with aqueducts, pipes, wells, and cisterns. The methods he suggested for conducting water were by artificial channels or within pipes of lead or baked clay. If a channel was used, it should have a very solid foundation and a minimum slope of 1 in 200. Vitruvius was probably concerned about losses of water by evaporation, and accordingly suggested that channels should be covered.

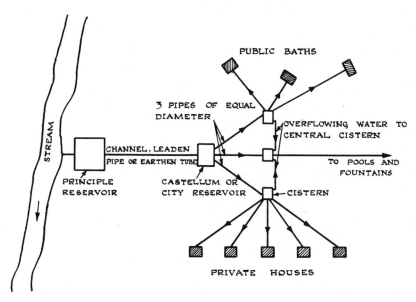

Figure 2. Water supply system of a Roman city.

The artificial covered channel brings water from the river to a reservoir built within the city walls. From there it is to be carried within three pipes to three interconnected tanks (figure 2). The first tank should supply water to the baths, the central one to the basins and fountains of the city, and the third to private houses. 'If the water is to be conducted in lead pipes, first build a reservoir at the source; then, let the pipes have an interior area corresponding to the amount of water, and lay these pipes from this reservoir to the reservoir which is inside the city walls. Pipes should be cast in lengths of ten feet.' He preferred clay pipes to lead ones because it would be comparatively easy to repair damages and because water conducted therein is 'not harmful but wholesome'.

Vitruvius, like other Romans, seemed totally unaware of the fact that discharge depended on both the velocity and cross-sectional area of a stream. (Hero of Alexandria was perhaps the only contemporary Roman who had a clearer conception of that phenomenon.) It was the general practice of the Romans, including Vitruvius, to evaluate discharge by measuring only the cross-sectional area of a stream or by measuring the area of an orifice or a conduit through which the water flowed. Vitruvius also erred about the source of the Nile. His contemporary, Juba II (d. 20 A.D.), the King of Mauretania, stated in his work *Libyca* that the source of the Nile was in Western Mauretania,[8] and probably Vitruvius was aware of his work. The Nile was a constant wonder to the people of the ancient civilizations and they kept speculating time and again about its origin and the regularity of its inundation.[8]

In the first century A.D., the Roman polymath Pliny quoted Vitruvius extensively, although nowhere did he acknowledge the source of that information.

MEASUREMENT OF DISCHARGE

Of all the varieties of hydrologic data, stream flow records are probably of the greatest importance to hydrologists. However, the present-day simple concept that discharge is equal to the cross-sectional area times the velocity $(Q = A \times V)$, which is almost the first lesson in elementary hydraulics, took a long time to make a permanent appearance. It was first enunciated by Hero of Alexandria but it seems to have attracted very little attention at that time. The principle was never used until 1628,

when it was derived independently by Benedetto Castelli.*
Considerable controversy exists amongst the historians of science
over the time in which Hero of Alexandria lived. The time varies
from about 150 B.C. to 250 A.D.; but the present consensus[9] seems
to be that Hero flourished sometime after 65 A.D. but before 150
A.D. If this is correct, Hero was post-Vitruvian and almost a
contemporary of the Roman water commissioner Sextus Julius
Frontinus. Two of the most important works of Hero, chief of the
Alexandrian School, are *Pneumatica* and *Dioptra*. Like Thales, he was
a practical man, and was mainly concerned with the practical
applications of knowledge. Consequently it is no surprise to find
that he has been sometimes called 'the first engineer'.[10]

In *Pneumatica* Hero described more than 20 methods of application
of syphon, and explained its use for both drainage and irrigation.
According to him, syphons were extensively used for irrigating
lands bordering the desert, and could be successfully used for
conveying water over across long hills and valleys.

The second book *Dioptra* was translated by Pappus in the third
century A.D. It is a remarkable book on land surveying, and is so
called because of the instrument described in it which could be used
for similar purposes as the modern-day theodolite. From the stand-
point of hydrology, the most important contribution of Hero is his
method of determining the discharge of a spring. The problem and
the solution as given by him is as follows:

'Given a spring, to determine its flow, that is, the quantity of water which it
delivers.
One must, however, note that the flow does not always remain the same.
Thus, when there are rains the flow is increased, for the water on the hills being
in excess is more violently squeezed out. But in times of dryness the flow subsides
because no additional supply of water comes to the spring. In the case of the
best springs, however, the amount of flow does not contract very much.
Now it is necessary to block in all the water of the spring so that none of it runs
off at any point, and to construct a lead pipe of rectangular cross-section. Care
should be taken to make the dimensions of the pipe considerably greater than
those of the stream of water. The pipe should then be inserted at a place such

---

* Leonardo da Vinci made studies of the distribution of velocities in open
channels, but the author has found no evidence yet that he actually computed
any discharge values from such data.

Figure 3. The first correct measurement of stream flow by Hero of Alexandria (reconstructed by Arthur H. Frazier).

that the water in the spring will flow out through it. That is, the pipe should be placed at a point below the spring so that it will receive the entire flow of water. Such a place below the spring will be determined by means of the dioptra. Now the water that flows through the pipe will cover a portion of the cross-section of the pipe at its mouth. Let this portion be, for example, 2 digits* [in height]. Now suppose that the width of the opening of the pipe is 6 digits, 6 × 2 = 12. Thus the flow of the spring is 12 [square] digits.

It is to be noted that in order to know how much water the spring supplies it does not suffice to find the area of the cross-section of the flow which in this case we say is 12 square digits. It is necessary also to find the speed of flow, for the swifter is the flow, the more water the spring supplies, and the slower it is, the less. One should therefore dig a reservoir under the stream and note with the help of a sundial how much water flows into the reservoir in a given time, and thus calculate how much will flow in a day [figure 3]. It is therefore unnecessary to measure the area of the cross-section of the stream. For the amount of water delivered will be clear from the measure of the time.'[12]

The concept that stream discharge is dependent on velocity, as suggested by Hero, was an isolated instance, and was not accepted

---

* The digit is an ancient Greek unit of length called *dactylos* and is equal to 1.85 cm.[11]

probably because he was much ahead of his time. As previously indicated, two great men of that period, Vitruvius and Frontinus, considered that the discharge amounted to the cross-sectional area of the stream, and theirs was the accepted practice as of that time. Since Vitruvius' concept has already been discussed, only Frontinus' approach will next be briefly described.

Sextus Julius Frontinus (35?–104 A.D.), one time Governor of Britain under Emperor Vespasian, received his appointment as commissioner of water works (*curator aquarum*) of Rome from Emperor Nerva in about 97 A.D. Though he was about sixty-two years old when he accepted this new job, he had a great love for the work. He probably began writing his masterpiece *De aquis urbis Romae, libri II* during the year that followed.[13] The book contains a wealth of information on water supply systems and methods used by the Romans during the first century A.D.

Herschel made a careful study of Frontinus' work, and of the water supply systems in use during the corresponding period. He believed that Frontinus' writings reflected the teachings of the Alexandrian school of mathematicians, especially those of Hero of Alexandria.[14] He seems to have assumed that Hero was at least older than the Roman water commissioner. If that assumption was correct (which is debatable), it would indicate that Frontinus obviously failed to grasp the fundamental law of flow in open channels or conduits as had been advocated by his teacher.

As has been explained, Frontinus had no clear concept of the fact that discharge in an open channel is dependent on velocity, cross-sectional area and slope. The Romans, like the Egyptians before them, were aware that in order to flow, water requires a downward slope. The slopes of the aqueducts were more closely related to the topographical conditions than to hydraulic considerations. The bottom slopes of the same aqueduct frequently varied considerably from about 1 in 2000 to 1 in 250. It is highly unlikely that Roman *mensors*, or *librators*, or *architectons* had any idea of reconciling a particular cross-sectional area with a definite slope in such a manner as to produce a desired discharge. All they seem to have done was to construct a part of the aqueduct, and if the resulting discharge was too little for their liking, they may have either increased the area (less likely) or just increased the slope (more probable). Thus, in all probability, the slope was fixed by a process of trial and error. Another belief in regard to his subject is that the Romans could have laid out or maintained any desired slope they wanted for their aqueducts with the aid of their favourite type of level, the *chorobate*. The erratic slopes of some reaches in the aqueducts have accordingly been ascribed to a lack of precision of either the *chorobates* or of the men who operated them.[15]

Despite the fact that he did nothing about it, Frontinus did have some vague ideas about the effect of head and velocity on the discharge. He stated, in paragraph 35, that:

'Let us not forget in this connection that every stream of water whenever it comes from a higher point and flows into a delivery tank through a short length of pipe, not only comes up to its measure but yields, moreover, a surplus; but whenever it comes from a low point, that is, under a less head, and is conducted a tolerably long distance, it will actually shrink in measure by the resistance of its own conduit; so that on these accounts, either an air or a check is needed for the discharge.'

Again in paragraph 33 he said:

'Whence it appears, that the amount measured by me is none too large; the explanation of this is, that the more impetuous stream of water increases the supply, since it comes from a large and rapidly flowing river.'

Since these two statements fall into a pattern, Frontinus probably believed that there is a *normal* or *standard* discharge for a particular cross-sectional area, and if the actual discharge exceeded the *standard* discharge allocated to the particular area, it is either due to un-usual velocity or higher head. But what was the *standard* discharge for a stream? The incriminating evidence comes from paragraph 65 where he speaks of Appia aqueduct:

'Appia is credited in the records with 841 quinariae. This aqueduct could not be gauged at the intake, because it there consists of two channels; but at the Twins, which is below Spes Vetus, where it joins the branch of the Augusta, I found a depth of water of 5 feet, and a width of one foot plus $^3/_4$, making $8^3/_4$ square feet of area, twenty-two 100-pipes [pipes of nominal area of at least 100 square digits] plus one 40-pipe, which makes 1825 quinariae; more than the records have it by 984 quinariae. It was discharging 704 quinariae; less than credited in the records by 137 quinariae, and, furthermore, less than given by the gauging at the Twins by 1121 quinariae.'

It can be clearly seen from the following calculations that Frontinus took discharge to be equal to the cross-sectional area.

Area = 5 × 1.75 = 8.75 sq. ft
             = 1260 sq. in.

1 *quinaria*   = 0.69026 sq. in. (area of an orifice $1^1/_4$ digits in diameter)

Discharge $= \dfrac{1260}{0.69} = 1825$ *quinariae*

The water commissioner was unable to balance his books, and it is not surprising because he completely disregarded velocities as well as slopes. However, he did not look far for such discrepancies. The reasons, he asserted, were due to leakages and to fraudulent practices of the Romans who seemed to be experts in tapping water without 'bothering' the authorities. Even allowing for leakage and illegal practices of the Romans, it can be pointed out that the apparent discrepancy was probably because the velocities in pipes were higher than in the feeders supplying the aqueduct.

Their established unit of measurement of flowing water was one *quinaria*. It was the area of a pipe of $1^1/_4$ *digits* in diameter. Herschel from his studies concluded that one *quinaria* was about '5000 or 6000 U.S. gallons per twenty-four hours, plus or minus 2000 or 3000 U.S. gallons, according to circumstances, favourable or un-favourable'.[16] It should, however, be pointed out that the equality of the rate of discharge with the cross-sectional area is permissible under certain circumstances. Frontinus stated that in 'setting ajutages [a short length of pipe of fixed diameter on which water charges were based], care must be taken to set them on a level, and not place the one higher and the other lower down. The lower one will take in more, the higher one will suck in less ...' If this condition was satisfied, the pipe marked 2 *quinariae* would then provide twice the discharge of the pipe of 1 *quinaria* according to the formula $Q = A \sqrt{2gh}$, assuming that the pipe is discharging into the air, and entrance and frictional losses in short lengths are neglected.

It is firmly believed that the concept of Frontinus for the measurement of discharge was the one that prevailed during the period of the Roman Civilization. The lone exception was that explained by

Figure 4. The aqueducts of Rome, restored. The Anio Novus and the Claudia in the left foreground, combined in a single structure; the Marcia, Tepula, and Julia also borne by one single arcade, on the right (painting by Zeno Diemer, by courtesy of Deutsches Museum, Munich).

Hero, whose correct understanding of the phenomenon unfortunately had gone unheeded.

AQUEDUCTS OF ROME

For nearly four and a half centuries after the foundation of the city, the Romans were using water either from the Tiber, or from wells and springs. They were indebted for their first aqueduct[17] to Appius Claudius Crassus, statesman, financier, and even a poet, who was responsible for building the Appia in 312 B.C. Building aqueducts was not a new art, as will become evident from a reading of chapter 1, but just the present-day ruins of the Roman aqueducts, with their magnificent structures, extensive systems of sluices and gates, long miles of canals, and pretentious outlooks arouse great admiration among modern engineers.[18] In the fifth century A.D., Rutilius Namatianus said: 'Why should I mention the aqueducts, sustained upon lofty arches, to which Iris could scarcely lift the waters of the clouds? One might say these were mountains that had grown star-ward ... Rivers are intercepted and hidden in thy walls. The lofty baths consume the whole of reservoirs'. Later Fabretti described them as *Romanae providentiae magnitudinisque primitiae* (the first fruits of Rome's foresight and greatness). Frontinus, who was in love with his waterworks, can well be pardoned for his statement: 'Will any-body compare the idle Pyramids, or those other useless though much renowned works of the Greeks, with these aqueducts, with these many indispensible structures?'

Details of the aqueducts of Rome[19] as they existed while under the jurisdiction of Frontinus are given in table 2, and figure 4 is the famous painting *Wasserleitungen in altem Rom* by Zeno Diemer in which appear the Anio Novus, the Claudia, the Marcia, the Tepula, and the Julia.

CONTRIBUTION OF SENECA

Lucius Annaeous Seneca (4 B.C.–65 A.D.), a Spaniard (figure 5) was brought to Rome by his father at an early age. He had the unenviable distinction of being associated with Caligula the mad-man (37–41 A.D.), Claudius the imbecile (41–54 A.D.), and Nero the monster (54–68 A.D.). He paid the penalty for having lived in such troubled times by being ordered to prepare for death by his former pupil Nero, who gallantly allowed him a choice of various means with which to perform the enforced suicide. The Stoic is best known for his moralistic writings and tragedies, but to hydro-logists his work *Quaestiones naturales* written soon after 63 A.D.,[20] is of major interest. That work consists of seven books which deal with astronomy, physics, physical geography, and meteorology. He draws heavily on Greek sources[21] – mainly Aristotle, Theophrastus, Posidonius, and Asclepiodotus. Unfortunately the meteorological writings of Posidonius and Asclepiodotus are no longer extant and only about four pages of an abstract of Theophrastus' work made by an unknown Arab has survived.[22] There is a great gap in the development of science from Aristotle and Theophrastus to Seneca but there is even a greater gap between Seneca and the beginning of the Renaissance.

TABLE 2

Aqueducts under Frontinus' charge.

Name	Builder	Date built	Source	Length (miles)	Size of* aqueduct (ft)	Quality of water	Elevation of delivery above Tiber wharves (ft)	Number of delivery tanks	Amount in quinariae		
									Available water	Used in city	Used outside the city
Appia	Claudius	312 B.C.	Spring	10.29	2.0 × 6.0	Excellent	28	20	704	699	5
Anio Vetus	Dentator	272–269 B.C.	River	39.55	2.5 × 7.0	Turbid	84	35	1610	1102	508
Marcia	Marcius	144–140 B.C.	Spring	56.73	4.6 × 9.0	Excellent	125	51	1935	1098	837
Tepula	Caepio and Longinus	125 B.C.	Spring	11.00	2.0 × 3.5	Warmest	128	14	445	331	114
Julia	Agrippa	33 B.C.	Spring	14.19	1.5 × 5.0	Excellent	133	17	803	597	206
Virgo	Agrippa	19 B.C.	Spring	12.97	2.2 × 5.0	Excellent	35	18	2504	2304	200
Alsietina	Augustus	10 A.D.	Lake	20.39	5.8 × 8.7	Not palatable	—	—	392	—	392
Claudia**	Caligula and Claudius	38–52 A.D.	Spring	43.34	3.0 × 6.2	Excellent	158	92	5625	3824	1801
Anio** Novus	Caligula and Claudius	38–52 A.D.	River	53.98	4.3 × 9.0	Turbid	158				
				262.44				247	14018	9955	4063

* The size of channels vary from place to place and hence dimensions are only approximate.

** The Anio Novus (upper) and the Claudia (lower) form a double-decked aqueduct.

Figure 5. Bronze head of Lucius Annaeus Seneca from Herculaneum (by courtesy of National Museum, Naples).

In Book IV of *Quaestiones naturales* Seneca discusses hail and snow in a rather frivolous manner. He said that he would be rather audacious if he suggested hail is formed in the sky exactly in the same way as ice on the earth, except in the previous case the whole cloud is frozen.[23] He then decided to imitate the chroniclers who, after lying to their heart's content, refuse to take responsibility for any particular statement and refer to the authorities for its authenticity. Hence, if his friends doubt his words about the formation of snow and hail, they can refer back to Posidonius who would vouch for it as if he had witnessed the whole process himself. Hail is formed from a rain-cloud which has just turned into liquid. It has a round shape because drops of all kinds tend to be globular, and even if it started its descent in an irregular shape, its sides would get rounded as it falls whirling through thick air. Snow, on the contrary, is not spherically shaped because it is not so solid and it does not fall from as great a height. Hail is simply ice suspended in mid-air, and similarly snow is suspended hoar-frost. He finishes his dissertation in a light vein, and then proceeds to inquire into the distribution of density and temperature in the atmosphere.

Seneca's views on the origin of rivers and springs will be discussed in the next subsection along with those held by some others of his period, and a discussion of his writings on the causes of the rise of the Nile will appear in chapter 6.

THE HYDROLOGIC CYCLE

The Latin poet and philosopher Lucretius Carus Titus (earlier half of the last century B.C., possibly 96?–55 B.C.) left a work *De rerum natura* (On the nature of things). It is his only work that is known to still exist.[24] He believed that moisture rises from everything, especially from the sea. When headlong winds drive the 'clouds' across a deep sea, the clouds pick up an abundant supply of moisture just like a woolen fleece soaks up dew. Then the upraised vapours assemble in a thousand ways only to lose their water content for either of these two reasons:

'The power of wind drives it [densely packed clouds] along.
The very multitude of clouds collected in a great array
And pushing from above,
Makes rains stream out in copious shower.
Then too,
When clouds are scattered by the winds or broken up,
Smitten above by rays of sun,
They send their moisture out and drip
As lighted tapers held above a scorching fire
Drip fast.'[25]

This process constitutes only one-half of the hydrologic cycle. The other half is discussed in connection with the sea,[26] which remains constant in size despite the large quantity of water being constantly added to it by rivers, springs, torrential rains, as well as by subterranean streams. Compared to the vastness of the sea, however, all the extra water coming in is hardly equal to a single drop. Besides, the sea loses a considerable amount of water due to evaporation from both the heat of the sun and winds sweeping the surface of the sea, and, as already explained, 'clouds' pick up much moisture. Since the earth is porous much water is lost from the sea by leakage, and, then:

'The brine is filtered off, the moisture trickles back
Assembles at the source of every stream
Thence flows through earth in torrent fresh
Wherever once a way is cut
To let the waters run in liquid march.'[26]

Thus Lucretius believed that rivers do not originate from precipitation but from the filtration of sea-water which is lost due to seepage. His concept on the origin and rise of the Nile[8] will be discussed in chapter 6.

From the presently available written works of antiquity, Vitruvius may be credited with the first pluvial explanation of the hydrologic cycle although, as discussed in the previous chapter, he probably got the idea from Greek sources – from either Theophrastus or Posidonius or from both of them.

Book III of Seneca's *Quaestiones naturales* is devoted to various forms of water but mainly to both surface and subterranean springs and rivers. He classifies water into two main categories: standing, as in lakes, and running, as in rivers and springs. Water is stagnant if it lies on a horizontal plane, and it is running if it exists on sloping ground. But why does not the sea grow larger from the water it continually receives from the rivers? And, as a matter of fact, why does not the earth run out of water from this continual loss? Seneca was ready with the answers but before presenting them he first stated the opinions expressed by some learned men. He refers to Lucretius' concept, but as it is not good enough to even be commented upon, he moves on to another one, namely that water in a river originally comes from precipitation. The philosopher strongly opposed that concept. As a diligent digger among his vines, he confidently asserted that even the heaviest rainfall could not percolate to a depth of more than 10 ft below the surface. When the ground was dry, all water is absorbed by the upper layer, and nothing reaches the lower one; and when it is saturated, the rain-water would find its way into river channels. River levels do not rise with first rainfall because the thirsty ground absorbs it all.

Next, he inquires, how can rivers originate from rocks and mountains? Rainfall has to flow off the bare crags because there is no earth to infiltrate. This surely must be an unanswerable question. In fact it is, but if the stoic had been observant enough, he would have failed to find a single spring issuing from any summit without its having any appreciable gathering grounds. Finally, as rainfall can not infiltrate to more than 10 ft below the surface, how can one explain the existence of rich springs of water at a depth of 200 to 300 ft in the driest of localities?

Having refuted the two main theories basic to the hydrologic cycle, he gave his own thoughts on the subject which, as can be seen from the following discussion, were undoubtedly coloured by Aristotle's teachings. Three reasons were suggested for the origin of underground water. They are:

(i) earth itself contains moisture which is forced out at the surface;

(ii) air within the earth is being continually converted into water by the underground forces of perpetual darkness, everlasting cold and inert density; and

(iii) the doctrine of interconvertibility of elements, i.e., earth within its interior turns itself into water.

If one is surprised at the large and continuous supply of fresh water in rivers, he has only to look at the size of reservoirs from where it came:

'Surely you might as well be surprised, when the winds drive hither and thither the whole atmosphere, that the supply of air does not fail, but flows on day and night increasingly. And the wind, remember, is not confined to a definite channel, as rivers are, but goes with wide sweep over the broad expanse of heaven. You might well, too, be surprised that after so many breakers have spent their force, any succeeding wave is left.'[27]

If one still asks how water is produced, he is met with a similar question: how air or earth is produced. There are four elements in nature, and one is not entitled to ask where water came from; it is just one of the four parts of nature.

Having settled the question, Seneca proceeded to explain why water in certain water courses stops flowing at times. The reason is simple

as the channels are cut off by rocks or earth displaced by earth-
quake.[28] He believed that the earth contains not only veins of water
but also subterranean rivers, huge lakes, and a hidden sea from all
of which the rivers at the surface obtain their supplies of water.[28]
Finally, every one knows that there are some standing waters which
have no bottoms, and this water is the perpetual source of large
rivers.

Seneca's discussion of the origin of the Nile will be treated with the
others in chapter 6.

One of the greatest legacies of Roman times preserved from antiquity is the *Natural history* of Pliny the Elder (Gaius Plinius Secundus,
23–79 A.D.). He was a prolific writer who had an indefatigable
curiosity about natural phenomena. He met his death while
recording an eruption of Mount Vesuvius, a martyr to scientific
curiosity. It is to be noted that the writings of Lucretius, Seneca
and Pliny on meteorological subjects have marked resemblances
and dissimilarities which probably indicate that they were drawing
their materials from the same primary sources.[29]

Pliny was confident that the freezing of rain is the cause of both
hail and snow; hail occurs when it is frozen hard, snow occurs when
it is less firmly concreted. Similarly, hoar-frost is frozen dew. As to
the origin of springs and rivers, he said:

'The earth opens her harbours, while the water pervades the whole earth,
within, without and above; its veins running in all directions, like connecting
links, and bursting out on even the highest ridges; where, forced up by air,
and pressed out by the weight of the earth, it shoots forth as from a pipe . . .'[30]

RAINFALL MEASUREMENTS AT PALESTINE

After Kautilya, the next mention of rainfall measurement appears
in a Palestinian book of religious writings called the *Mishnah*.
The book records nearly 400 years of Jewish cultural and religious
activities in Palestine, from sometime around the earlier half of the
second century B.C. to the close of the second century A.D.[31] In
a doctoral dissertation, Vogelstein[32] has pointed out that rain
gauges were used in Palestine during the time of the *Mishnah*.[33-36]
Rainfall was recorded during an entire year, and the total time was
divided into three periods:

'First, that of the early rain, which moistens the land and fits it for the reception of the seed, and is consequently the signal for the commencement of ploughing. Second, the copious winter rain, which saturates the earth, fills the cisterns and pools, and replenishes the springs. Third, the latter or spring rain, which causes the ears of corn to enlarge, enables the wheat and barley to support the dry heat of the early summer, and without which the harvest fails ... As a rule, it may be considered that the autumn or early rains extend from the commencement of the rainy season in October or November until the middle of December, the winter rains from the middle of December until the middle or end of March, and the latter or spring rains from the middle of March until the termination of the rainy season in April or May.'[37]

The rainfall values recorded for corresponding periods were as follows:

Early rain period	1 *tefah* (about 9 cm)
Second period	2 *tefahs*
Third period	3 *tefahs*
Annual rainfall	6 *tefahs* = 54 cm.

It is not possible to find out if the results were isolated readings for one year or averages for a number of years. It was also said that rainfall percolated to a depth of 1 *tefah* in barren soils, 2 *tefahs* in medium soil, and 3 *tefahs* in broken up arable lands. Obviously like the Indian practice, measurements of rainfall in Palestine were initiated primarily because of its importance to agriculture. Generally benedictions were said at the beginning of the rain period as well as after a certain amount of rain had fallen (when a container having a volume of 0.137 litre had been filled).[32]

It is rather interesting to note the amount of rainfall then considered as normal for a good harvest corresponds quite closely with the observations made by Thomas Chaplin at Jerusalem during the late nineteenth century.[38]

CONCLUSION

The major development during the Roman Civilization, so far as the science of hydrology is concerned, was undoubtedly the concept of Hero of the measurement of discharge, but because Hero was far ahead of his time, it is no surprise that his idea attracted scant attention. Both Vitruvius and Frontinus believed that discharge of a natural stream or discharge through a pipe is equal to its cross-sectional area, irrespective of its velocity. It took another seventeen centuries for the father and son team of John and Daniel Bernoulli to separately publish in 1738, their different mathematical demonstrations of the elementary principle of flow; $v = \sqrt{2gh}$. The Romans were aware that area of a pipe is equivalent to $1/4D^2$ but they seem to have had no concept of cubic measures, and units of discharge, like cubic feet per second, were perhaps beyond their wildest imagination.

Romans were practical people. While reading *De aquis* one gets the impression that Frontinus realized that something was wrong with his concept of equating area with discharge. Thus he speaks vaguely of the effects of velocity, head of water, boundary resistance, and frictional loss without having clear conception of any of these phenomena. The Romans built great water works and quite justifiably they were proud of them. It is concluded that they were built by purely empirical methods, without much understanding of physical principles, but it should be emphasized that they did work and served their purposes admirably.

So far as the concept of the origin of rivers and springs is concerned, it was stated reasonably well by Vitruvius. It is highly likely that the pluvial concept was first given by Theophrastus, but in the absence of any authentic records it will always remain a purely conjectural assumption.

Finally, the quantitative measurement of precipitation by the Jews of Palestine for agricultural purposes was an important development. There seems to be no connection between the Indian practice of the fourth century B.C. and the Palestinian development of the first century A.D. Both were independent and isolated practices which did not continue for a long time.

REFERENCES

1. SARTON, G., A history of science. Hellenistic science and culture in the last three centuries B.C. Cambridge, Harvard University Press (1959) pp. 350–360.
2. CLAGETT, M., Greek science in Antiquity. London, Abelard–Schuman Ltd. (1957) pp. 104–107.

3. STAHL, W. H., Roman science. Madison, The University of Wisconsin Press (1962) pp. 92–96.
4. VITRUVIUS, Ten books on architecture, translated by M. H. Morgan, book 1. Cambridge, Harvard University Press (1914) p. 13.
5. *Ibid.*, p. 226.
6. VITRUVIUS, The architecture of Marcus Vitruvius Pollio in 10 books, translated by J. Gwilt, book 8. London, Priestly and Weale (1926) pp. 234–235.
7. KIRCHER, A., Mundus subterraneus. Amstelodami, Apud J. Janssonium et E. Weyerstraten (1664–1665).
8. BISWAS, ASIT K., The Nile: its origin and rise. Water and Sewage Works *113* (1966) 282–292.
9. NEUGEBAUER, O., Über eine Methode zur Distanzbestimmung Alexandria-Rom bei Heron. Det Kgl. Danske Videnskabernes Selskab. Historisk-filologiske Meddelelser, vol 26, no. 2. København, Levin and Munksgaard (1938).
10. TRUESDELL, W. A., The first engineer. Journal of the Association of Engineering Societies *19* (1897) 1–19.
11. ARGYROPOULOS, P. A., Contribution of the ancient Greeks to present day hydraulic knowledge. Bulletin 81, Engineering Experiment Station. Lexington, University of Kentucky (1966) pp. 11–16.
12. COHEN, M. R. and I. E. DRABKIN, A source book in Greek science. Cambridge, Harvard University Press (1948) p. 241.
13. GROSSMAN, E., Frontinus, the water commissioner. Journal of the Boston Society of Civil Engineers *16* (1929) 417–420.
14. HERSCHEL, C., Frontinus and the water supply system of the city of Rome. Boston, Dana Estes & Co. (1899) p. 106.
15. D'AVIGDOR, E. H., Water works, ancient and modern. Engineering *21* (1876) 500–501.
16. HERSCHEL, C., *op. cit.*, p. 215.
17. VAN DEMAN, E. B., The building of Roman aqueducts. Publication no. 423, Carnegie Institute of Washington (1934).
18. BELL, T. J., History of the water supply of the world. Cincinnati, Peter G. Thomson (1882).
19. DRAFFIN, J. O., The story of man's quest for water. Champaign, Ill., The Garrard Press (1939).
20. CLAGETT, M., *op. cit.*, pp. 108–110.
21. STAHL, W. H., *op. cit.*, pp. 98–100.
22. LUCRETIUS, De rerum natura, edited by C. Bailey, vol. 3. Oxford, University Press (1947) pp. 1745–1748.
23. SENECA, Physical science in the time of Nero, translated by J. Clarke. London, Macmillan and Co. Ltd. (1910) pp. 177–179.
24. BISWAS, ASIT K., The hydrologic cycle. Civil Engineering, ASCE *35* (1966) 70–74.
25. LUCRETIUS CARUS, TITUS, De rerum natura, translated by A. D. Winspear. New York, S. A. Russel (1956) p. 268.

26. *Ibid.*, pp. 273–274.
27. SENECA, *op cit.*, pp. 120–121.
28. *Ibid.*, pp. 233, 235.
29. BISWAS, ASIT K., Hydrology during the Roman Civilization. Water and Sewage Works *114* (1967) 344–347, 373–376, 422–425.
30. PLINY, The natural history of Pliny, translated by J. Bostock and H. T. Riley, vol. 1, ch. 66. London, Henry G. Bohn (1855) p. 97.
31. THE MISHNAH, translated by Herbert Danby. Oxford, University Press (1933).
32. VOGELSTEIN, H., Die Landwirthschaft in Palästina zur Zeit der Mišnah. Part I, Getreidebau. Dissertation. Breslau (1894).
33. HELLMANN, G., Die Entwicklung der meteorologischen Beobachtungen bis zum Ende des XVII. Jahrhunderts. Meteorologische Zeitschrift (1901) 145–157.
34. HANN, J., Die ältesten Ragenmessungen in Palästina. Meteorologische Zeitschrift (1895) p. 136.
35. ANONYMOUS, The invention of raingage. Scientific American *103* (1910) 504.
36. HORTON, R. E., The measurement of rainfall and snow. Journal of the New England Water Works Association *33* (1919) 14–23.
37. CHAPLIN, T., The early and latter rain. Jews and Christians *2* (1895) 69–89.
38. HELLMANN, G., The dawn of meteorology. Quarterly Journal of the Royal Meteorological Society *24* (1908) 221–232.

6

The origin and the rise of the river Nile

INTRODUCTION

'Egypt is an acquired country, the gift of the River' (*doran tu potamu*), so said Herodotus, the father of history. Many ancient philosophers asserted that the Nile was superior to the other rivers of the world – that it was in a class by itself. Very few physical facts of antiquity received more discussion than the annual inundation of the river Nile. The Greek geographer and historian Strabo stated of the river that 'its rising, and its mouths were considered, as they are at the present day, amongst the most remarkable, the most wonderful, and most worthy of recording of all peculiarities of Egypt'.[1] Greek philosophers were so intrigued by the regularity of the floods of the Nile that some believed that the river had been created along with the world, and that reason only could explain its peculiar characteristics. It was considered to be too vast and remarkable to have had the same origin as other rivers, and that is why one reason was sometimes advanced for its origin and another for the other rivers of the world.[2] The present chapter deals exclusively with the various theories put forward on the origin and rise of the special and almost 'legendary' river, the Nile.

ETESIAN WINDS OF THALES

Thales of Miletos, the chief of the seven wise men of ancient Greece,

seems not to have handed down any works of his own. All our
knowledge concerning him comes from the writings of other
philosophers (see chapter 2). Thales conjectured that northerly
etesian winds blowing against the direction of the flow of the Nile
prevented the water from running into the sea.[3-6] Etesian is the
northerly wind which blows during summer with reasonable regularity. Herodotus discounted the theory because the Nile had risen
even during years when the etesian wind had failed to blow. Moreover, if that were the cause, why are rivers in Syria and Libya,
which also flow into the wind, not affected in a similar manner –
particularly when they are smaller rivers with weaker currents.
It is interesting to note Herodotus's differentiation between strong
and weak currents. He, perhaps, was the first man to do so.

The Spaniard Seneca strongly criticized Thales for his theory, and
Euthymenes of Marseilles for his having supplied corroborative
testimony. Apparently Euthymenes sailed down the Nile to what he
claimed was the Atlantic Ocean, and observed that the rise in
the Nile coincided with the etesian winds. (He probably lived
towards the end of the sixth century B.C. – a time at which if he
were questioned as to how the Nile was connected with the Atlantic
Ocean, he himself would undoubtedly have failed to provide a
satisfactory answer.) Nevertheless, Seneca had a variety of reasons
for opposing Thales' theory. One was that the rise of the Nile does
not coincide with the blowing of the wind. That wind starts before
the rise, and lasts until well after the river subsides. The rise also does
not vary in unison with the blasts of the wind. 'Then, again, the
Etesian winds beat on the shore of Egypt, and the Nile comes down
in their teeth; whereas, if its rise is to be traced to them, the river
ought to come from the same quarter as they do. Furthermore, if it
flowed out of the sea, its waters would be clear and dark blue, not
muddy, as they are.'[3,4] From the last objection it may be deduced
that Seneca did not completely understand Thales' theory. Seneca
obviously did not think much of Euthymenes' testimony as 'in
those days there was room for lying; when there was no knowledge
of foreign parts, it was easy for foreign parts to ship us romances'.[7]

Figure 1. The world according to Hecataios of Miletos.

'OCEANUS' CONCEPT

The primitive Greek geographers imagined that the earth is en-
circled by an immense and swift stream called Oceanus. It existed
so far beyond the sea that there was no mixing between the waters.
It has no source or outlet. From it rose all the stars, excepting those
of the constellation of the Great Bear, only to plunge back again.
Herodotus stated that some were of the opinion that the Nile flows
from Oceanus, and gave that reason for explaining its peculiar
characteristics (figure 1).
How the concept of Oceanus came into being is extremely difficult

to imagine. Homer had used it in both the Odyssey and the Iliad. In fact, he probably made the problem more complicated by interweaving mythology with geography therein. The legend of Oceanus is a charming story.[8] Titan Oceanus was the son of Uranus (heaven) and Gaea (earth), and was considered to be one of the basic elemental forces responsible for the creation of the world. He was an extremely powerful God, so much so that Homer considered his power second only to that of Zeus. He was wedded to his sister Tethys, and by her had three thousand rivers and three thousand Oceanids (sea-nymphs) – quite a remarkable achievement. He thus became the father of all the rivers, seas, and other types of primeval waters.

MELTING SNOWS OF ANAXAGORAS

Anaxagoras of Clazomenae went to Athens immediately after the Persian Wars. According to him the rise of the Nile is due to the melting of snow on the peaks of the Lybian mountains where the river begins. His explanation, though rational, is not entirely correct. Nevertheless, Aeschylus, Sophocles, and Euripides shared his view. Herodotus rejected Anaxagoras' view because it is 'positively farthest from truth'. His argument was that wind blowing from the direction of Libya is extremely hot. He believed that there must be rain within five days of snowfall, and since rain and frost are unknown in that country, how could snow fall at all? Swallows and kites remain there throughout the year, and cranes migrate there to escape the Scythian winter.[5, 6] All this tends to prove that the reason suggested is an impossible one. Though Herodotus was completely wrong in his supposition, one must give him credit for displaying coherence of thought in his search for truth.

Seneca said that the country is so hot that the Troglodytes (cave dwellers) built underground houses,[3, 4] and silver became unsoldered. He admits there is some snow in the Ethiopian mountains, but if it were actually the cause, the Nile would rise in late spring or early summer as do the rivers originating in the Alps, Thrace, or Caucasus where it snows heavily. Melting occurs quickly with fresh and lightly packed snow. Old and hard packed snow melts more slowly. Thus early summer runoff should far exceed that which occurs later. The Nile floods however last for four full months, and its rate of rise is uniform – hence, the Anaxagorean theory has to be discounted.

ORIGIN OF EXPLANATIONS

Herodotus mentioned all three of the above theories, but following his usual practice, he fails to name their originators. Diels in his monumental work *Doxographi Graeci*[9] has shown that all the three theories originated from the pseudo-Aristotelian treatise *On the rise of the Nile*. The writer attributed the first theory to Thales of Miletos, the second to Euthymenes of Marseilles, and the third to Anaxagoras. The question at once arises as to where the pseudo-Aristotelean author got those names. Probably, from Hecataios; as Herodotus was familiar with that geographer's views (in fact he often copied them). Hecataios is probably the earliest author who had even referred to that otherwise obscure person, Euthymenes.

EXPLANATION OF HERODOTUS

The Ionian, Herodotus, who considered all knowledge to be within his dominion,[10] thought he knew the sources of all rivers except the Borysthenes (present Dnieper) and the Nile, but he received a little comfort for this ignorance by saying 'Nor, I think does any Greek'. He wanted to find the reasons for (a) the Nile's regular annual inundation, (b) its behaviour – which is just the reverse of other rivers, and (c) its inability to create a breeze. One must admire the historian's curiosity in his earnest attempts to unravel the causes of these physical oddities. Apparently the Egyptians had no theories thereon, nor what is more important, did they even try to explain them. They were content with their faith in the Nile god, Hapi, and were willing to take anything and everything for granted. Hapi was depicted as a fat, bearded man with full breasts from which gushed the life-giving water (figure 2). He was dressed like a boatman and a fisherman, and wore a crown made from a sheaf of lotus plants. The Egyptians believed that the Nile had two entities: one, the Nile of Egypt, the other, the celestial Nile which flows across the

Figure 2. Hapi, the Nile god (by courtesy of Trustees of the British Museum, London).

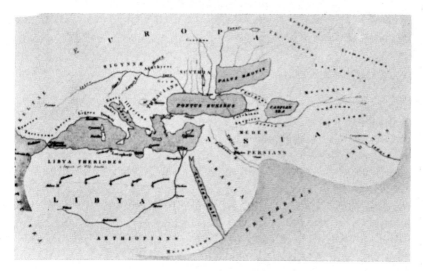

Figure 3. The world of Herodotus.

heavens and can be seen as a luminous river (Milky Way). Herodotus rejected the etesian wind, melting snow, or the Oceanus
theories as being true explanations for the rise of the Nile. He
considered that they were presented by the Greeks for the sole purpose of advertising their own cleverness. He thereupon proposed
an astounding hypothesis of his own, namely:

'The sun, when he traverses the upper parts of Libya, does what he commonly
does in summer – he draws the water to him and having thus drawn it, he pushes
it to the upper regions [of the air probably] and then the winds take it and
disperse it till they dissolve in moisture. And thus the winds which blow from
those countries, Libs and Notus, are the most moist of all winds. Now when the
winter relaxes and the sun returns to the north, he still draws water from all
the rivers, but they are increased by showers and rain-torrents, so that they are
in flood till the summer comes; and then, the rain failing and the sun still
drawing them, they become small. But the Nile, not being fed by rains, yet
being drawn by the sun, is, alone of all rivers, much scanty in the winter than in
the summer. For in summer it is drawn like all other rivers, but in winter it
alone has its supplies shut up.'[11]

Herodotus believed that the upper Nile flows in the same direction
as Danube – west to east (figure 3). He also confused the Niger with
the Nile. But considering the fact that such erroneous ideas continued

to exist in one form or another, for the next 2200 years, perhaps the historian deserves to be excused.

OENOPIDES OF CHIOS

Oenopides of Chios, who was in his prime around the third quarter of the fifth century B.C., was a contemporary of Hippocrates. His theory was the most popular one[12] in ancient times. He maintained that during the winter, the internal heat stored in the ground dries up all the underground veins of water, and rivers continued to flow only because of the rainfall they received. Since the Nile valley does not experience precipitation, its flow must gradually dwindle. The internal heat also caused caves to be warm in winter as well as the water in deep wells. But as the summer approaches, the heat disappears, and as a consequence the water will flow back into the Nile and cause floods.

DIOGENES OF APPOLONIA

Diogenes of Appolonia was a Greek eclectic philospher of the fifth century B.C. He was interested in physiology and cosmology, and was responsible for the reconciliation of the doctrines of Anaximenes of Miletos and Anaxagoras of Clazomenae. He believed that the water loss caused by the withdrawal of moisture by the sun is counterbalanced by the earth by drawing extra water – partly from the sea and partly from another source. During the summer the southern part of the earth becomes parched as the heat of the sun affects it most. The earth is interconnected by numerous secret channels, and through them water from the wet zone comes to the drier because wet and dry can not exist together in nature. Thus 'just as in a lamp, the oil flows to the point where it is consumed, so the water inclines toward the place to which the overpowering heat of the burning earth draws it'.[13] The water thus attracted comes from the superabundant source of the northern region of eternal winter. Consequently, there is a flood of water travelling in one direction. If this process did not occur, the whole earth would have either dried up or flooded a long time ago. It was also the reason ascribed for the continuous swift current from the Black Sea to the Lower Sea in contradistinction to the alternate flow and ebb of tides in other seas. Seneca flatly opposed the theory. He said:

'Now, one would like to ask Diogenes, seeing the deep and all streams are in inter-communication, why the rivers are not everywhere large in summer. Egypt, he will perhaps tell me, is more baked by the sun, and therefore the Nile rises higher from the extra supply it draws; but in the other countries, too, the rivers received some addition. Another question – seeing that every land attracts moisture from other regions, and a greater supply in proportion to its heat, why is any part of the world without moisture? Another – why is the Nile fresh if its water comes from the sea? No river has a sweeter taste.'[13, 4]

SNOW AND ETESIAN WINDS OF DEMOCRITUS

Democritus of Abdera (460?–357? B.C.) was perhaps the greatest of all physical philosophers. He claimed to have 'wandered over a larger part of the world than any other man of my time, inquiring about things most remote; I have observed very many climates and lands and have listened to very many learned men'.[14] He declared that the snow melts and flows away in the northern parts during the summer solstice thus forming clouds by the vapours. The etesian wind drives the clouds towards the south and Egypt, and gives rise to violent storms which fill up the lakes and the Nile.

His theory is very interesting on two counts. It hints that the Nile has its source lakes in Central Africa and that they are rain-fed lakes. Even more important is his concept of the movement of storm systems, since until the eighteenth century it was commonly believed that storms did not move from one place to another.

EPHORUS TO STRABO

The Greek historian Ephorus (400–330 B.C.) believed that 'all Egypt, being porous and made of river silt, and formed like pumice stone has long continuous crannies, and through these it takes up a great quantity of moisture, which it contains within itself in winter time, and in summer emits on all sides as it were streams of sweat; and it is through these that the river fills'.[15]

There is considerable controversy over Aristotle's opinion on the Nile. Sarton[16, 17] considers it to be post-Eratosthenian, but Partsch[18] believes that it had been written either by Aristotle himself or by one of his contemporaries. Whoever did write it claimed that the heavy rainfall during the spring and the early summer in the highlands of the Blue and the White Nile was responsible for the creation of floods in the lower river.

Eratosthenes (276–194/192 B.C.), the chief librarian at Alexandria, drew a reasonably accurate map of the Nile up as far as what is now Khartoum, and hinted that the equatorial lakes are the sources of the river. He said that there is no reason why man should speculate on the rise of the Nile, because those sources of the Nile have been explored, and that heavy rainfalls have been observed there. Eratosthenes also mentioned that Aristotle had previously suggested this same theory.

Strabo of Amasya was a geographer who considered Homer to be the source of all knowledge and wisdom. He ridiculed Herodotus as being a 'marvel-monger'. He traced the authorities for the rise of the Nile through a series of philosophers back to the 'Master' – Homer.

'For he [Poseidonius] says, that Callisthenes asserts that the cause of the rise of the river [Nile] is the rain of summer. This he borrows from Aristotle, who borrowed it from Thrasyalces the Thasian [one of the ancient writers on physics], Thrasyalces from some other person, and he from Homer who calls the Nile "heaven-descended": "Back to Egypt's heaven-descended stream".'[19]

Strabo quoted another philosopher, Nearchus, as saying that the rises of the Nile and the Indian rivers are caused by summer rains.[20] When Alexander the Great saw crocodiles and Egyptian beans during his conquest of north-western India, he thought that he had discovered a river that extended to the source of the Nile. He was about to equip a fleet to sail up that river to Egypt. Apparently he soon changed his mind, and Strabo soon traced his reason therefore back to the works of his master – Homer!

In the Middle Ages the Austrian Benedictine monk, Engelbertus Admontansis wrote a commentary about the pseudo-Aristotelean treatise on the flooding of the Nile.[21]

LUCRETIUS TO BEDE

The didactic epic *De rerum natura* (on the nature of things) was written by perhaps the greatest Roman poet, Lucretius Carus Titus

(96?–45 B.C.). According to him, several causes must be cited for the many strange things in nature. It is like coming across a dead man who could have died by violence, poison, disease, or from cold. One of them must be the actual reason – but which one, one does not know. This, be suggested, is the case with the Nile. He stated the reasons put forward by Thales, Anaxagoras, and Democritus and added the following one (probably his own):

'It may be too
That heaps of sand pile up against the river mouths
And check the current of the stream,
When sea stirred up by heavy winds drives sand within,
And so it comes about
The exit of the water is less free
The waves flow down less easily.'[22]

The King of Mauretania, Juba II (d. 20 A.D.), a historian and archaeologist, was the author of the work *Libyca*, from which Pliny has quoted. He revived an old theory about the course of the Nile but gave it a new form (figure 4). He stated that the source of the Nile is in western Mauretania – not far from the Ocean. From there it travels underground for a several days' journey to a similar lake

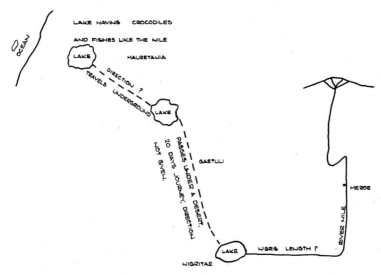

Figure 4. The course of the Nile according to Juba II, the King of Mauretania.

in Mauretania Caesaiensis; thence underground again for another twenty days' journey (directions are not given) to the source Nigris – at the borderline between Africa and Ethiopia. From there it continues under the name Astapus, through Ethiopia. This absurd theory was accepted by many. It is probably the origin of the erroneous but long-lived idea that the Niger is a branch of the Nile. In one form or another, eminent scholars like Pliny, Mela, Vitruvius, and Strabo were later to become subscribers to this error.

The treatise *Questiones naturales* by the Spaniard Lucius Annaeous Seneca mentions the views of other ancient philosophers on the rise of the Nile. His objections to them have already been described. The discussion thereon ends so abruptly that it is obvious that either it is incomplete or else there is a lacuna in the text. Seneca tends to agree with the ancient philosophers who stated that certain rivers were originally created as part of the world, and that this circumstance alone can explain their peculiar characteristics. He believed that the Nile and the Danube were too remarkable to have had origins of the same natures as those of other rivers. Unfortunately, he did not give any specific reason for the regularity of the inundation of the Nile. He does, however, mention that Nero dispatched two centurions up the Nile to find its source. They travelled upstream until they came to a limitless marshy region which was so covered with vegetation that they found it impossible to proceed further either by boat or by foot. In that region they found 'two rocks from which an enormous body of the river came out'. Seneca probably believed that the water of the Nile came out of the earth, and such a belief would be compatible with his concept of origin of underground water (see chapter 5).

The Roman savant, Pliny the Elder (Gaius Plinius Secundus), is renowned for his book on natural history. He thought that the two most probable theories were those of Thales and Democritus. He also noted a theory promulgated by Timaeus, the mathematician, who:

'alleged a reason of an occult nature: he says that the source of the river is known by the name of Phiala, and that the stream buries itself in channels underground where it sends forth vapours generated by the heat among the steaming rocks amid which it conceals itself; but that, during the days of inundation, in consequence of the sun coming closer to the earth, the waters are drawn forth by influence of his heat, and that on being thus exposed to the air, it overflowed; after that, in order that it may not be dried up completely, the stream hides itself again. He says that this takes place at the rising of the Dog-star, when the sun enters the sign of Leo, and stands in a vertical position over the source of the river, at which time no shadows occur at that spot.'[23]

After Pliny, the English theologian and historian Venerable Bede (674–735 A.D.) compiled a summary of the causes of the inundation of the Nile.[24]

During the fifteenth century, Leonardo da Vinci remarked that the source of the Nile can be traced to three very high lakes in Ethiopia.

'It issues forth from the Mountains of the Moon from diverse and unknown beginnings; and comes upon the said lakes high above the watery sphere at an altitude of about four thousand braccia, that is a mile and a third, in order to allow for the Nile to fall a braccia in every mile.'[25]

CONCLUSION

The various theories as well as the objections put forward against them by the ancient philosophers have been discussed in this chapter. Probably one of the greatest indirect benefits of the annual inundation of the Nile is the science of geometry. Its discovery arose from the need to make new land measurements after every flood. To quote Diadochus:

'For the Egyptians had to perform such measurements because the overflow of the Nile would cause the boundary of each person's land to disappear. Furthermore, it should occasion no surprise that the discovery both of this science (geometry) and of other sciences proceed from utility . . . And so, just as accurate knowledge of numbers originated with the Phoenicians through their commerce and their business transactions, so geometry was discovered by the Egyptians for the reason we have indicated.'[26]

Another theory about the annual rise of the Nile was proposed during the seventeenth century. The theory, which received an unexpected amount of support, and which will be discussed later, was that the floods were caused by the fermentation of nitre.

REFERENCES

1. STRABO, The geography of Strabo, book 1, translated by H. C. Hamilton and W. Falconer. Bohn's Classical Library. London, Henry G. Bohn (1757) p. 46.

2. BISWAS, ASIT K., The Nile: its origin and rise. Water and Sewage Works *113* (1966) 282–292.

3. SENECA, Physical science in the time of Nero (*Questiones naturales*), translated by J. Clarke. London, Macmillan & Co. (1910) pp. 172–177.

4. COHEN, M. R. and I. E. DRABKIN, A source book in Greek science. Cambridge, Harvard University Press (1948) pp. 379–394.

5. HERODOTUS, The histories, translated by Aubrey de Selincourt. Harmondsworth, Penguin Books (1959) pp. 109–111.

6. HERODOTUS, The history of Herodotus, translated by G. Rawlinson. New York, Tudor Publishing Co. (1960) pp. 87–89.

7. ALEXANDER, W. H., Seneca's Quaestiones naturales: the text amended and explained. Publications in Classical Philology. Berkeley, University of California *13* (1948) 241–332.

8. LAROUSSE ENCYCLOPAEDIA OF MYTHOLOGY. London, Batchworth Press Ltd. (1959) p. 167.

9. DIELS, H., Doxographi Graeci. Berlin (1879) pp. 226–229.

10. BISWAS, ASIT K., Experiments on atmospheric evaporation till the end of the 18th century. Technology and Culture *10* (1969) 49–58.

11. WHEWELL, W., History of the inductive science, from the earliest to present time, vol. 1. London, John W. Parker and Son (1857) pp. 24–25.

12. FORBES, R. J., Studies in ancient technology, vol. 7. Leiden, E. J. Brill (1963) p. 22.

13. SENECA, *op. cit.*, p. 176.

14. FRISINGER, H. H., Early theories on the Nile floods. Weather *20* (1965) 206–207.

15. WARMINGTON, E. H., Greek geography. London, J. M. Dent and Sons Ltd. (1934) p. 42.

16. SARTON, G., Introduction to the history of science, from Homer to Omar Khayyam, vol. 1. Baltimore, Williams and Wilkins Co. (1927) pp. 135-136.

17. SARTON, G., A history of science, ancient science through the Golden Age of Greece. Cambridge, Harvard University Press (1953) p. 559.

18. PARTSCH, J., Das Aristotles Buch 'Über das Steigen des Nil'. Abhandlungen der Philologisch–Historischen Klasse der Königlichen Sachsischen Gesellschaft der Wissenschaften. Leipzig, B. G. Teubner (1909) pp. 553–600.

19. STRABO, *op. cit.*, p. 225.

20. *Ibid.*, p. 88.

21. SARTON, G., Introduction of history of science, vol. 3, part 2. Baltimore, Williams and Wilkins Co. (1947) p. 1845.

22. LUCRETIUS CARUS, TITUS, De rerum natura, translated by A. D. Winspear and S. A. Russel. New York, The Harbor Press (1955) p. 278.

23. PLINY THE ELDER, The natural history of Pliny, translated by John Bostock and H. T. Riley. Bohn's Classical Library. London, Henry C. Bohn (1855).

24. BEDE, VENERABILIS, The miscellaneous works of Venerable Bede in the original Latin, vol. 6. Opuscula scientifica. London, Whittaker and Co. (1843).

25. MACCURDY, E., The notebooks of Leonardo da Vinci, vol. 2, new ed. London, Jonathan Cape (1956) p. 129.

26. COHEN, M. R. and I. E. DRABKIN, *op. cit.*, p. 34.

7

From 200 to 1500 A.D.

INTRODUCTION

The dawn of the third century of the Christian era revealed an extraordinary mixture of different religions, sects, and philosophies at Alexandria, the centre of learning. The Christian, Judaic, and pagan religions were not on friendly terms; the Alexandrian school as well as the schools of Plato and Aristotle were decaying fast; and the presence of Stoics and Epicureans tended to increase the difficulties. But there was one factor that was common to all elements of this curious conglomeration – it was a profound contempt for science.[1]

HYDROLOGY UNDER THE HOLY FATHERS

With the advent of power acquired by the Christian Church, the desire as well as the privilege of investigating natural phenomena gradually faded away, and with the exception of theology a great stagnation spread over all branches of knowledge. Knowledge was permissible only as a means of edifying the doctrines of the scripture, as interpreted by the Holy Fathers. The Christian faith became a matter of eschatology and honest criticism gradually disappeared. The supreme function of knowledge was the furtherance of salvation and the highest task of philosopher was still to use his knowledge in the service of theology, and to demonstrate philosophy's handmaidenly accord with revealed Christian truth.[2] According to Dampier:

'The Roman empire died, but its soul lived on in the Catholic Church, which took over its framework and its universalist ideals ... Philosophically the Catholic Church was the last creative achievement of Hellenistic civilisation; politically and organically it was the offspring and heir to the autocratic Roman Empire.'[3]

Isidore of Seville

At the turn of the seventh century, when the desecularization process started by the Catholic Church in its determined effort to eradicate heresies and heathenism all over Western Europe was complete, there lived in Seville a Spanish bishop and scholar, Saint Isidore (570–636). He can favourably be compared with Pliny, and his approach to science was similarly encyclopedic. His twenty-volume work on *Etymologies* or *Origins* is not as systematic as the Roman's *Natural history*, but it is to be remembered that during Isidore's time science and scholarship were at their lowest ebb; and powerful members of the Church, like Pope Gregory, were outspoken opponents of all secular knowledge. When considered in that light, *Etymologies* was an outstanding accomplishment, and it was one of the most widely read books for the next one thousand years.[4] When looked at from any direction, that was no small achievement.

Isidore explained meteorological phenomena primarily from a basis of the four elements.[5] Like Aristotle, he also believed in the interconvertibility of air into water, or water into air.[6] He gave two different reasons for precipitation.

'Air being contracted, makes clouds; being thickened, rain; when the clouds freeze, snow; when thick clouds freeze in a more disordered way, hail; being spread abroad, it causes fine weather, for it is well-known that thick air is cloud, and a rarified and spread-out cloud is air.'[7]

Later on in his book he said:

'Rains (*pluviae*) are so called because they flow, as if *fluviae*. They arise by exhalation from earth and sea, and being carried aloft they fall in drops on the lands, being acted upon by the heat of the sun or condensed by strong winds.'[8]

These concepts probably indicate that Isidore was aware of the meteorological works of the Greek and the Roman philosophers, perhaps through intermediaries.

Chapters 12 and 13 of Book XIII of *Etymologies* deal with water and their different qualities from distinctly theological viewpoints. Isidore described water as the most powerful element because,

'the waters temper the heavens, fertilize the earth, incorporate air in their exhalations, climb aloft and claim the heavens, for what is more marvelous than the waters keeping their place in the heavens.'[9]

He then goes on to discuss the various springs and lakes of the world, especially their magical and therapeutical properties. In later chapters he described the oceans, the Mediterranean Sea (the first recorded use of the word as a proper name), gulfs of the world, tides and straits, lakes and pools, the great subterranean abyss of water, various rivers of the earth, and finally the floods which, needless to say, were biblical, starting with Noah's.

Isidore believed that there is a huge abyss under the earth from which all the springs and rivers originate. The abyss is bottomless and all waters eventually return there through secret channels. This is almost a restatement of the Platonic concept of Tartarus but it is more likely to have come from the book of *Ecclesiastes*, chapter 1, verse 7, which reads:

'All rivers run into the sea;
Yet the sea is not full;
Unto the place from whence the rivers come,
Thither they return again.'

Isidore assigned various reasons for the elevation of the sea remaining constant. He claimed it was because:

'its very greatness does not feel the waters flowing in; secondly, because the bitter water consumes the fresh that is added, or that the clouds draw up much water to themselves, or that the winds carry it off, and the sun partly dries it up; lastly, because the water leaks through certain secret holes in the earth, and turns and runs back to the source of rivers and to the springs.'[10]

Venerable Bede

Venerable Bede (673–735), like Isidore, was a devout Churchman and was interested in knowledge insofar as it is useful in propagating a Christian point of view. Intellectually Bede was superior to Isidore, and along with Boethius (480?–524?) he may be ranked as one of the two top intellectual figures of the Latin West during the early Middle Ages. He was the first Englishman to write about weather, and has sometimes been called the founder of English meteorology.[11] Bede was not an original observer of nature, nor was he a theorizer; but he was a compiler and extraordinary summarizer of the existing knowledge from classical sources. His book *De rerum natura* served as an elementary primer on Christian cosmography and astronomy for monks. He freely borrowed from the book *Natural history* of Pliny and even appropriated an entire chapter from it. He acknowledged the fact, however, and even referred his readers to Pliny for additional information.

His outlook on hydro-meteorology was traditional. For example, wind is the air in motion or disturbance of air; as can be proved with a fan.[12] Breaking of the clouds produces thunder; hail melts more rapidly than snow, and it falls more often during the day than during the night. Salt in the sea cannot be raised by the sun's rays, and hence sea-water is saline; whereas rain, rivers or lakes are not salty. It may be noted that sun's rays which lack the power to lift the salt from the sea, are reported to be capable of stopping planets in their courses! He also proposed a theory about the origin of the Nile[13] which was a combination of the ideas of Thales and Lucretius. He said that the etesian winds are responsible for causing the sea waves to deposit sands at the mouths of the Nile. As the sands pile up, the exits become less free with the consequent rise of the water level.[14]

HYDRO-METEOROLOGY

After Bede there was no significant development in the field of hydrology until the time of the Italian master, Leonardo da Vinci, but that should not be taken to mean that there was no interest in hydro-meteorology. As a matter of fact, Hellman has shown[15] that there were twenty-six works dealing substantially with some aspects of that subject between the seventh and the fourteenth centuries (although they were just dreary variations of previous works with the probable exceptions of a tenth century encyclopaedia written in Basra by a secret society called 'The Brethern of Purity').[16] The vague discussion it contained about the role of cooling during the process of precipitation follows:

'If the air is warm, these vapours [vapour from the sea and heated exhalation from the land] rise to a great height, and the clouds collect one above the other stepwise, as it is observed in spring and autumn. It is as if they were mountains of combed cotton, one over another. But if cold from the zone of ice comes in from above, the vapours collect and become water; then their parts are pressed together, and they become drops, increase in weight and fall from the upper region of the cloud down through its mass. These little drops unite with one another until, if they come out of the lower boundary of the cloud, they are large drops of rain. If they meet great cold on their way, they freeze together and become hail before they reach the ground. In consequence, those that come from the upper part of the cloud will be hail, but those from the lower boundary of the cloud will be rain mixed with hail ... So the lower boundary of the region of icy cold, and the high mountains round about the sea confine the two rising streams of vapour from which clouds and rain come; they scatter them and take them away, just like the walls and roofs of bath-houses.'[17]

Toward the end of the twelfth century, works of Aristotle (including *Meteorologica*) were again available to the western world through Latin translations made in Spain[18] from the Arabic versions thereof. The meteorological ideas of the great master were in vogue again with the scholars and universities, under the subject title of *Meteora*. The spirit of inquiry was rekindled, as will be obvious from the following questionnaire handed down to us by a Scotsman, Michael Scot (1175?–1232):

'Likewise tell us how it happens that the waters of the sea are sweet although they all come from the living sea. Tell us too concerning the sweet waters how they continually gush forth from the earth ... where they have their source and how it is that certain waters come forth sweet and fresh, some clear, others turbid, others thick and gummy; for we greatly wonder at these things, knowing already that all waters come from the sea, which is the bed and receptacle of all running waters. Hence we should like to know whether there is one place by itself which has sweet water only and one with salt water only, or whether there is one place for both kinds, ... and how the running waters in all parts of the world seem to pour forth of their superabundance continually from their source, and although their flow is copious yet they do not increase if more were added beyond the common measure but remain constant at a flow which is uniform or nearly so.'[19]

These questions went unanswered (in the sense correctly) for about half a millennium in spite of Scot's promise to answer them all. In the thirteenth century Albertus Magnus (1206–1280) wrote two famous treatises entitled *De meteoris* and *De passionibus aeris* which relied heavily on the works of Aristotle. Then, or slightly later, Vincent de Beauvais in France, Thomas de Cantimpre in Belgium, Ristoro d'Arezzo in Italy, and Bartholomew Anglicus in England, wrote about physical science incorporating mainly Aristotelean ideas, all under the very convenient title, *De rerum natura*. Probably the first book in the English language dealing with the hydro-meteorological subjects was published in 1481. It was called *Mirrour of the world*,[20] and was an English version of the work *Image du monde* of the thirteenth century as 'translated out of frensche into English by me simple person Wyllm Caxton'. The book, reprinted in 1912, deals with clouds, rain, snow, frost, hail, tempest, the circulation of water and salinity of the sea – the same subjects written about so much by the Greek, Roman and the Medieval writers.

The Italian, Leone Battista Alberti (1404–1472), was said to have 'made so great a progress in the sciences that he outstript all the great men of that age who were most famous for their learning'.[21] A major portion of Book X of his treatise *Ten books on architecture*, is devoted to water, and it reminds one of the 8th book of Vitruvius' *Architecture in 10 books*. Alberti was much influenced by Vitruvius but he was better-read than his Roman counterpart. He summarized the properties of water as follows:

'We must not omit to take notice of what we see with our eyes, that water naturally tends downwards; that it cannot suffer the air to be anywhere beneath it; that it hates all mixtures with any body that is either lighter or heavier than itself; that it loves to fill up every concavity into which it runs; that the more you endeavour to force it the more obstinately it strives against you, nor is it ever satisfied till it obtains the rest which it desires, and then when it is got to its place of repose, it is contented only with itself, and despises all other mixtures; lastly, that its surface is always an exact level.'[22]

On the question of the origin of springs and rivers, Alberti discussed all the explanations that had previously been offered, and concluded that he could not pretend to determine the correct answer because there are 'so much variety among authors upon the subject, and so many different considerations offering themselves to the mind when we think upon it'.[22] But he thought that if somebody believed that rivers originated from rainwater he will 'not be much mistaken in his conjecture'. He had some ideas about ground water table, and said that if a well is dug, water will be found only if it is sunk to the level of the nearest river.

RAIN GAUGES IN CHINA AND KOREA

The problem of flooding of rivers and canals has always been serious in China and, hence, it is not surprising to find that rain gauges were used at least as early as 1247 A.D. It is to be noted that the previous precipitation measurements were made in Palestine some twelve centuries ago[23] (see chapter 5), but that quantitative hydrological observations were completely lacking during a substantial part of the intervening period.

The book *Shu Shu Chiu Chang* (Mathematical treatise in nine sections) by Chhin Chiu-Shao has a series of problems concerning the shapes of rain gauges, called *thien chhih tshe yü*. The rain gauges he described were of conical- or barrel-shaped vessels. There was one installed at every provincial and district capital. The book also deals with problems on snow gauges – *chu chhi yen hsüeh*. Those gauges, consisting of large cages made of bamboo, were placed at the sides of mountain passes and uplands. This is probably the very first use of snow gauges on record. Chiu-Shao also discussed a method of determining the amount of rainfall over an area from observations of point rainfalls.[24]

The need for regular rainfall during rice cultivation periods probably resulted in the introduction from China, of the practice of obtaining rainfall measurements all over Korea during the fifteenth century. The very economy of the country depended on the annual production of rice; so much so, in fact, that from ancient times the Koreans prayed for rain to the saints of the mountains and rivers.

The first mention of rain gauges appeared during the reign of King Sejong of the Lee Dynasty in 1441 A.D.[25, 26] during the Golden Age of the Korean Civilization. Rain gauges of the type then described (figure 1) were used throughout Korea until 1907 A.D. and later references to them occur frequently in the annals of Korean history. The earliest one reads:

'In the twenty-fourth year of the reign of King Sejo, the King caused an instrument of bronze to be constructed for measurement of rainfall. It was a vase

Figure 1. Korean rain gauge of 1441 A.D. (by courtesy of Dr. C. Cook, Director Central Meteorological Office, Seoul).

1 *shaku*, 5 *sun* deep and 7 *sun* broad, set on a pillar. The instrument was placed at the observatory and the officials of the observatory measured the depth of rainfall each time it rained. The results were made known to the King. Similar instruments were distributed to the provinces and the cantons, and the results of the observations were reported to the court.'[27]

The Korean rain gauges were about 30 cm deep and 15 cm in diameter. Early in the twentieth century, Wada, then the director of the Korean Meteorological Observatory, made a systematic search for those early rain gauges and for records of the observations made with them, but was unsuccessful.

Some 228 years after the time of King Sejong, rain gauges were mentioned as follows in books pertaining to Korean history:

'In the year 46 of the King Eijo [1770 A.D.], the King, following the ancient system of King Sejo, had numerous rain gauges constructed, and placed them, two at the Palace, by the side of the wind-vane, and the others in the chief places of the eight provinces. The rain gauges are placed on stone pillars, measuring 1 in height and 0.8 at the side, and on the upper surface is a hole 1 deep to receive the end of the instrument.'[28]

Wada did succeed in finding two rain gauges of this period – one in Seoul and the other at Taiku. The second one (figure 2) was still in actual use at the Observatory of Chemulpo.[29] On figure 2, the large three Chinese letters mean 'instrument to measure rain' and the smaller seven characters explain that it was 'constructed in the fifth month of the cycle of the year', during a date in the Chinese calendar which corresponds to 1770 A.D. Wada also found a cubical block of marble with a hole in the upper surface for holding a rain gauge, in the park of the Bureau of Registers at the Palace of the residence of H.M. Emperor of Korea. Some 360 Chinese letters are engraved in the pedestal saying that rain gauges were 'invented' during the reign of King Sejo and were later improved by King Eijo.

Unfortunately the report by Wada concerning the Korean rain gauges contained three errors. Such gauges were in use in Korea at least from 1441 A.D.[25, 26] and not 1442 A.D. The rain gauges were not first discovered by the Italian Benedetto Castelli in 1639 A.D., as supposed by him, but they were in use in India some two millenniums ago. Finally, like the majority of Korean astronomical

Figure 2. The rain gauge found at Taiku, 1770 A.D. (after Wada).

Figure 3. Plaster copy of an early Korean rain gauge at the Science Museum, London (Crown copyright).

instruments, the rain gauges were either imported from China or copied from the Chinese types.

Figure 3 shows a plaster copy of an early (1837 A.D.) Korean *tsche yu chhi* (rain-measuring instrument) now in the Science Museum, London.[30] Gauges such as this stood on *yun kuan thai* (cloud-watching platforms). The original rain gauge from which it was copied was the one at the Korean Observatory at Chemulpo described previously. Thus, the rain gauge of King Sejo was first copied in 1770 A.D. and recopied in 1837 A.D.

CONCLUSION

The science of hydrology did not advance much during the thirteen hundred years covered in this chapter. The scholars of this period believed or were led to believe that:

'Science remains subordinate to theology, and the statements of Scripture represent equally unassailable truths in science as in theology, that they, too, are bent on treating all passages in the Bible in which something is said about nature as scientific enunciations and bringing them into agreement with the results of professional science, does not prevent there being a difference of nuance between their conception of the study of nature and that of the Church Fathers.'[31]

It is, indeed, fortunate that man's inborn thirst for knowledge could not be suppressed by the admonition that its gratification was not conducive to his salvation. Hydrology, like any other science, suffered very badly from this intellectual suppression, but on the other hand, the general influence of people like Roger Bacon *(argumentum non sufficit, sed experienta)* must have indirectly affected the development of this particular branch of knowledge. In retrospect, the greatest achievement in the field of hydrology during these long thirteen hundred years were undoubtedly the redevelopment of quantitative measurement of precipitation in China and Korea.

REFERENCES

1. SINGER, C., Short history of scientific ideas to 1900. Oxford, Clarendon Press (1959) pp. 132–133.
2. TAYLOR, H. O., Thoughts and expression in the 16th century, vol. 2. New York, Macmillan & Co. (1920) p. 267.

3. DAMPIER, W. C., A history of science, 4th ed. Cambridge, University Press (1961) pp. 66–67.

4. STAHL, W. H., Roman science. Madison, University of Wisconsin Press (1962) pp. 214–215.

5. BREHAUT, E., An encyclopedist of dark ages: Isidore of Seville. Studies in history, economics, and public law, vol. 48 New York, Columbia University (1912) p. 60.

6. Ibid., pp. 235–242.

7. Ibid., p. 237.

8. Ibid., p. 238.

9. Ibid., p. 239.

10. Ibid., p. 240.

11. BOTLEY, C. M., A founder of English meteorology. Quarterly Journal of the Royal Meteorological Society 61 (1935) 346.

12. BEDE, VENERABILIS, The miscellaneous works of Venerable Bede in the original Latin, vol. 6. Opuscula scientifica. London, Whittaker and Co. (1843).

13. BISWAS, ASIT K., The Nile: its origin and rise. Water and Sewage Works 113 (1966) 282–292.

14. FRISINGER, H. H., Early theories on the Nile floods. Weather 20 (1965) 206–207.

15. HELLMAN, G., Denkmäler mittelalterlicher Meteorologie. Neudrücke von Schriften und Karten über Meteorologie und Erdmagnetismus, vol. 15. Berlin (1904).

16. SARTON, G., Introduction to the history of science, vol. 1. Baltimore, Williams and Wilkins Co. (1927) pp. 660–661.

17. Quoted by W. E. K. MIDDLETON, A history of the theories of rain. London, Oldbourne Book Co. Ltd. (1965) pp. 15–16.

18. HELLMAN, G., The dawn of meteorology. Quarterly Journal of the Royal Meteorological Society 34 (1908) 221–231.

19. HASKINS, C. H., Studies in the history of medieval science. New York, Frederick Ungar Publishing Co. (1960) pp. 266–267.

20. CAXTON, J., Mirrour of the world, edited by O. H. Prior, Kegan, Paul, Trench, Trubner & Co., Ltd., Clarendon Press, Oxford, (1912).

21. DE FRENSE, R., The life of Leone Battista Alberti. In: Ten books of architecture, translated by J. Leoni. London, Alec Tiranti Ltd. (1955) pp. XII–XVI.

22. ALBERTI, L. B., Ten books on architecture, translated by J. Leoni. London, Alec Tiranti Ltd. (1955) pp. 213–214.

23. BISWAS, ASIT K., Development of rain gauges. Journal of Irrigation and Drainage Division, ASCE 93 (1967) 99–124.

24. NEEDHAM, J., Science and civilization in China, vol. 3. Cambridge, University Press (1959) pp. 471–472.

25. COOK, CHAEPYO, Personal communication (1966).

26. ANONYMOUS, Seoul monthly precipitation records 1770–1960. Seoul, Republic of Korea, Central Meteorological Office.

27. WADA, Y., Scientific memoirs of the Korean Meteorological Observatory, vol. 1. Chemulpo (1910). Quoted from Annales Historiques, vol. 2.

28. ANONYMOUS, Korean rain gauges of the fifteenth century. Quarterly Journal of the Royal Meteorological Society 37 (1911) 83–86.

29. HORTON, R. E., The measurement of rainfall and snow. Journal of the New England Water Works Association 33 (1919) 14–23.

30. LYONS, H. G., An early Korean rain gauge. Quarterly Journal of the Royal Meteorology Society 50 (1924) 26.

31. DIJKSTERHUIS, E. J., The mechanization of the world picture, translated by C. Dikshoorn. Oxford, Clarendon Press (1961) p. 100.

8

The sixteenth century

INTRODUCTION

During the later Middle Ages in general, only the people who desired to join the learned professions of Church, law, and medicine were educated at the universities,[1] and the entire intellectual attainment of Europe rested primarily on the scholars. An average man was contented if he was considered to be a good Christian (meaning that he was capable of reading the Bible and that he worried about saving his soul). The medieval scholar believed that many religious truths, like miracles, should remain unreasonable and inexplicable. When he asked himself questions such as 'from whence comes the water of a spring?' he first referred to the Scriptures. If he did not find it there, he might have looked into the works of the Greek or the Roman philosophers or even into those of his Islamic predecessors for the answers. But he did not, in general, venture to propose explanations of his own for such natural phenomena and, consequently, it is not surprising that the sciences of hydrology and meteorology did not advance much during those ages.

The first indications of a divorce between science and philosophy can be traced to about the beginning of the twelfth century, although a final split did not come until a few centuries later. The word philosophy included any type of inquiry, scientific or philosophical, using the roots of the modern terms. The Church had dominated the secular knowledge for such a long time that the impetus of individual independent thoughts like those of Roger

Bacon or Leonardo were unable to produce any significant effect on the status of science. It is true that the refreshing breeze of both the renaissance and the reformation brought an acceptance of new ideas again, but the writings of the old Greek masters kept on being dissected and summarized with monotonous regularity. It was still the same old wine, but with a new label on the bottle. With very few exceptions, people preferred book learning to observations of nature, and universities concentrated on classical literature. So far as experimental science was concerned, universities were as intolerant of it as was the contemporary Catholic Church. Such were the unfavourable conditions concerning the advancement of knowledge when the great Italian genius, Leonardo da Vinci, was born.

LEONARDO DA VINCI

Leonardo da Vinci (figure 1), illegitimate son of a Florentine lawyer, was born in 1452 in the small Tuscan village of Vinci, in Italy. In 1481, he wrote an extraordinary letter offering his services to Ludovico Sforza, the Duke of Milan, a copy of which is still in existence, though not in his own handwriting. In the letter he proclaimed himself to be an expert military engineer, architect, civil engineer, and mentioned that as a sculptor or painter 'I can do as well as anybody else, no matter who he may be'. The letter achieved its objective[2] and it would probably be true to say that he far surpassed his every claim. He died at Amboise, France, in 1519.

Leonardo was a genius, and few people, if any, will disagree with that statement. According to Benvenuto Cellini, Francis I believed that no other man 'had attained so great a knowledge as Leonardo, not only as a sculptor, painter and architect, but beyond that, as a profound philosopher'. Some have claimed that Leonardo even anticipated the discoveries of Francis Bacon, James Watt, Isaac Newton, and William Harvey, but that could be stretching the cult of the Florentine too far. At the other extreme, it has been said that Leonardo's techniques show very little originality, and that his works largely reflect those of Francesco di Giorgio (1439–1500) and other engineers.[3] Obviously, the truth lies somewhere between those two extremes. Be that as it may, he could have given hydrology

Figure 1. Leonardo da Vinci (by courtesy of Uffizi Gallery, Florence).

a great boost along the road of progress. The fact that he did not do so was largely his own fault since he devoted so much of his time to gathering new data that there was none left for preparing his notes for publication. He started accumulating those notes at the age of about 37, and continued to do so until almost the time of his death. During these years he filled over 7000 sheets with valuable notes and sketches on scientific subjects. They were not only condensed, but were written in 'mirror-image' handwriting, a method that seems natural for numerous left-handed individuals. No doubt these circumstances added to the problems involved in getting them in condition for publication, and although facilities were then available for doing so, none of his notes ever reached a publication stage during his life time. Leonardo, however, was aware of his short-comings, and in one of his manuscripts now at the British Museum, he wrote:

'This a collection without order, drawn from many papers that I have here copied, hoping later to put them in their right order, according to the subjects which they treat. I fear that I must repeat myself frequently; do not blame me for this, reader, because the subjects are numerous and the memory is unable to have them all present and say: 'I shall not write this because I wrote it before.'[4]

At the time of his death, the entire collection of notes was bequeathed to his young friend and companion, Francesco Melzi, who screened them for whatever they may have contained about Art, and largely ignored the others. Those 'discards' were eventually tied into between one and two dozen bundles, and were either given or sold to an equal number of libraries or individuals. And even after that took place, they continued to receive very little public attention. Eventually, some of them came to the attention of an outstanding hydraulic engineer, Giovanni Batista Venturi (1746–1822), the man for whom Clemens Herschel named the modern Venturi meter. After he had examined the notes, Venturi wrote an article entitled *Essai sur les ouvrages physico-mathématiques de Leonardo da Vinci* which was published in 1797. It revealed what a truly modern scientist Leonardo was as of that date, but more importantly, it stimulated the production of a still-continuing deluge of books and articles about him. Of particular interest here is the fact that by far the largest number of Leonardo's notes were devoted to hydrology and hydraulics than to any other single subject.

Leonardo on the hydrologic cycle

Leonardo's concepts on the hydrologic cycle have never received a thorough examination, and this is not surprising as only two serious studies have been made so far thereon. Both Adams[5, 6] and Meinzer[7, 8], who conducted them, concluded erroneously that Leonardo had a correct understanding of the concept of the hydrologic cycle. In his book on *The birth and development of geological sciences*, Adams reported:

'Leonardo da Vinci . . . was one of the earliest to see the true explanation of the origin of rivers. He wrote but little on the subject, yet from his observations in the Alps he recognized the important role played by the more pervious beds in the synclinal folds of great mountain ranges, especially when they are dipping at a high angle and lie between impervious beds, in carrying the rain and snow waters deep down into the crust of the earth, whence they may be brought again to the surface at some distant point or perhaps onward into the ocean without again reaching the surface at all.'[9]

Adam's statement was based on a secondary source of Leonardo's work by De Lorenzo.[10] Later, Meinzer quoted a passage from the Richter-translation of Leonardo's notes to support his contention that the Italian had a correct concept of the hydrologic cycle.[8] Unfortunately, the lines quoted, if studied in their proper context, reveal Leonardo's error. The statements of Adams and Meinzer have been widely quoted by subsequent writers. Not all of them shared those opinions, however. For example, Krynine, who has been mentioned previously in this work, never did credit Leonardo with having expressed the correct concept.[11] Be that as it may, a fairly thorough search of Leonardo's actual notes indicate that during the course of a long period of years, Leonardo did make a few clear and currently acceptable statements regarding the nature of the hydrologic cycle. A brief explanation of the circumstances follows.

From his familiarity with the writings of Pliny (23–79 A.D.) and Galen (130–200 A.D.), the founder of experimental physiology, it may be assumed that they might have had an appreciable amount of influence on Leonardo's thoughts, especially on the hydrologic cycle, as will soon become evident. Pliny, in his *Natural history* (II, 24, 66), stated:

'Water penetrates the earth everywhere, inside, outside, above, along connecting veins running in all directions, and breaks through to the highest mountain summits – there it gushes as in siphons, driven by pneuma (spiritus) and forced out by the weight of the earth; it would seem that the water is never in danger of falling; on the contrary, it bursts through to high places and summits. Hence, it is clear why the seas never grow from the daily influx of river water.'

Galen's influence was of a different nature. In following his example, Leonardo dissected numerous cadavers and searched for similarities. not only between structural make-up of man as compared with other living creatures on the earth, but also between man and inanimate objects such as the world itself. This is shown in the following thoughts he had prepared for use in his planned (but never completed) *Treatise on water*:

'By the ancients man has been called the world in miniature; and certainly this name is well bestowed, because, inasmuch as man is composed of earth, water, air, and fire, his body resembles that of the earth; and as man has in him bones, the supports and framework of his flesh, the world has its rocks, the supports of the earth; as man has in him a pool of blood in which the lungs rise and fall in breathing, so the body of earth has its ocean tide which likewise rises and falls every six hours, as if the world breathed; as in that pool of blood veins have their origin, which ramify all over the human body, so likewise the ocean sea fills the body of the earth with infinite springs of water. The body of the earth lacks sinews, and this is because the sinews are made expressly for movements and the world being perpetually stable, no movement takes place, and, no movement taking place, muscles are not necessary. But in all other points they are much alike … if the body of the earth were not like that of man, it would be impossible that the waters of the sea – being so much lower than the mountains – could by by their nature rise up to the summits of these mountains. Hence it is to be believed that the same cause which keeps the blood at the top of the head in man keeps the water at the summits of the mountains.'[12]

He carried the thought a step further:

'The same cause which moves the humours in every species of animate bodies against the natural law of gravity also propels the water through the veins of the earth wherein it is enclosed and distributes it through small passages. And as the blood rises from below and pours out through the broken veins of the forehead, as the water rises from the lowest part of the vine to the branches that are cut, so from the lowest depth of the sea the water rises to the summits of mountains, where, finding the veins broken, it pours out and returns to the bottom of the sea. Thus the movement of the water inside and outside varies in turn, now it is compelled to rise, then it descends in natural freedom. Thus joined together it goes round and round in continuous rotation, hither and thither from above and from below, it never rests in quiet, not from its course, but from its nature.'[13]

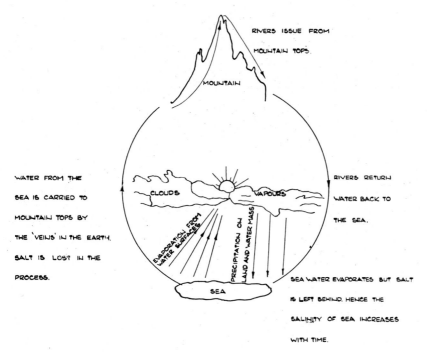

Figure 2. Leonardo's concept of the hydrologic cycle.

While the foregoing translations from Leonardo's own statements may seem to have little resemblance to the modern concept of the hydrologic cycle, it is important to consider that on the very same sheet which contained the longer thereof, Leonardo made the following statement:

'Where there is life there is heat, and where vital heat is, there is movement of vapour. This is proved inasmuch as we see that the element of fire by its heat always draws to itself damp vapours and thick mists as opaque clouds, which it raises from seas as well as lakes and rivers and damp valleys; and these being drawn by degrees as far as the cold region, the first portion stops, because heat and moisture cannot exist with cold and dryness; and where the first portion stops the rest settle, and thus one portion after another being added, thick and dark clouds are formed. They are often wafted about and borne by the winds from one region to another, where by their density they become so heavy that they fall in thick rain; and if the heat of the sun is added to the power of the element of fire, the clouds are drawn up higher still and find a greater degree of cold, in which they form ice and fall in storms of hail. Now the same heat which holds up so great a weight of water as is seen to rain from the clouds, draws them from below upwards, from the foot of the mountains, and leads and holds them within the summits of the mountains, and these, finding some fissure, issue continuously and cause rivers.'[14]

Here then, on the same page of notes, are two seemingly divergent concepts concerning the hydrologic cycle – one correct, the other in error, as viewed on the basis of our present ideas on the subject. Characteristically, Leonardo reported an occasional doubt about certain aspects of both theories, but nothing has been found so far which would indicate that he had at any time discarded the basic concepts of either of them. In fact, the chances seem good that he believed both systems operated concurrently. One can easily imagine Leonardo observing a mountain stream (which he considered a ruptured vein) in the side of the mountain during a sudden shower, with the water falling from the clouds supplementing the flow of the spring. The final paragraph in the above quotations seems to contain evidence that he was trying to express that very thought.

In view of what has been said here, it is admitted that not all of Leonardo's conception of the hydrologic cycle was in agreement with that which is currently held, but it is nevertheless contended that he should not, for that reason, be denied the right to receive full credit for having clearly and correctly expressed the basic feature of the modern concept thereof.[16] Figure 2 is a diagrammatic representation of such a concept.

Leonardo on open channel flow

Leonardo had a better understanding of the principles of flow in open channels than any of his predecessors or contemporaries, and it is not surprising as his ideas were based on observations. He was interested, ever since his youth, in the study of the various physical characteristics of mountains, rivers and seas. Figure 3, for example, is a photograph of two conjugate pages from Leonardo's works, one of which shows a philosopher* contemplating, and the other is several studies of whirling water. Except for a few isolated instances, the observations were made on the spot, and by himself. His

---

* The philosopher depicted could have been a self-portrait of Leonardo, but it is not certain because Leonardo was not so old at the time of the drawing.

Figure 3. Two conjugate pages from Leonardo's notebooks. One shows a philosopher (Leonardo?) contemplating, and the other shows a series of sketches of swirling water.

characteristic attitude was that to draw valid conclusions, one must rely on experience and experiments.

Fortunately for hydrologists and hydraulicians, various scholars have re-arranged the writings of Leonardo on water from his many manuscripts, notably Arconati[17] (1643), Cardinali[18] (1826), Mac-Curdy[15] (1928) and Richter[13] (1939), the last two being English translations.

Leonardo has discussed flow in open channels many times, and had obviously a clear concept of the principle of continuity, so much so that it could be named after him with complete justification. Consider the following passage:

'Given two rivers of equal volume of water at their entrances, their exits will be equal; that is, given an equal volume of water in an equal time, even though rivers may vary in length, breadth, slant and depth and the one be twisted and the other straight; or though both be twisted but the shapes of their curves are unlike; or one be of uniform breadth and the other of varying breadth; and if both vary their variation may be different; one may be of uniform depth and the

other of varying depth; and should both depths vary in themselves their variation
may not have any kind of likeness; and the whole of one may be uniformly
swift and the other uniformly slow, or the slowness and swiftness of one may be
mixed . . .; and the fact that there exist in these two rivers infinite varieties of
currents in breadth, length, slant and depth will not therefore prevent the equal
entrances of the one and the other from being equal in their exits.'[19]

He offered the now familiar analogy of the principle:

'Let us imagine an avenue formed of three consecutive sections, each of a different width; the first section, the narrowest [figure 4] is 4 times less wide than
the second, and this one [the second] is twice narrower than the third; people,
closely pressed against each other, fill these avenues; they must march together
in a continuous manner; when the people in the large avenue take one step
forward, those in the middle one must make two and those in the smallest,
eight; a proportion which you will find in all motions passing through sections
of width.'[20]

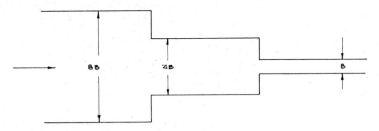

Figure 4. Continuity analogy of Leonardo.

Leonardo was aware of the major part played by slope in open
channel hydraulics:

'where the channel of the river is more sloping the water has a swifter current;
and where the water is swifter it wears the bed of its river more away and deepens
it more and causes the same quantity of water to occupy less space.'[21]

Velocity of flow. Consider the following statements of Leonardo:

'Waters which fall from the same level with an equal slant in an equal length
of movement will be of equal swiftness.'

'Of waters which fall from the same level by channels of equal slant, that will
have the swifter course which has the greater length.'

'Of waters which fall the same distance from the same level, that will have the
swifter course which has the greater length.'

'Of waters which fall the same distance from the same level, that will be slower which is longer.'

The concept was obviously erroneous because there is a direct functional relationship between velocity and distance. Had he introduced a time element instead of the distance, then undoubtedly he would have postulated one of the fundamental laws of motion (velocity = acceleration × time) nearly a hundred years before Galileo. However, he did realize that the accelerated motion cannot continue indefinitely because of side and bottom resistances, and he also had an excellent concept of velocity distribution.

'Water has higher speed on the surface than at the bottom. This happens because water on the surface borders on air which is of little resistance, because lighter than water, and water at the bottom is touching the earth which is of higher resistance, because heavier than water and not moving. From this follows that the part which is more distant from the bottom has less resistance than that below.'[22]

Figure 5 is an artist's conception of how Leonardo conducted his studies on velocity distribution in rivers. The rod floats near his

Figure 5. Measurement of velocity by Leonardo (reconstructed by Arthur H. Frazier).

helper are from Manuscript 'A', folio 42 verso, Institut de France. His sketch of the float is as shown in his note-book. Leonardo is keeping abreast of the float in the mid-river as it moves downstream. While doing so, he measures the distance traveled by means of his odometer; and he measures the time in which it travels that distance by singing musical scales up and down eight to ten times.[23] The other floats, held together by a piece of string instead of a wooden rod, and lying behind his helper, are from Codice Leicester, folio 13 verso. The surveying level and pitcher (for filling the groove cut into the top surface of the level) are from Codice 'B', 2173, folio 65 verso, Institut de France. Leonardo used a level, probably like the one shown, to determine the slope of the river. This is probably the first serious attempt to determine the velocity of flowing water by means of floats.

Treatise on water
Leonardo planned to write a treatise on water and even went to the extent of dividing it into fifteen books. The classification was as follows:

Book 1 of water in itself,
Book 2 of the sea,
Book 3 of subterranean rivers,
Book 4 of rivers,
Book 5 of the nature of the depths,
Book 6 of the obstacles,
Book 7 of different kinds of gravel,
Book 8 of the surface water,
Book 9 of the things that move in it,
Book 10 of the repairing (the banks) of rivers,
Book 11 of conduits,
Book 12 of canals,
Book 13 of machines turned by water,
Book 14 of raising water,
Book 15 of matters worn by water.

Even though his intention was evident, he never did complete the work. His manuscripts, especially Codex Leicester, contain a vast array of notes, figures, as well as brief outlines for chapters for the proposed work. One such note, dealing with precipitation, for example, reads:

'Write how clouds are formed and how they dissolve, and what it is that causes vapour to rise from the water of the earth into the air, and the cause of mists and of the air becoming thickened, and why it appears more blue or less blue at one time than another. Write in the same way of the regions of the air and the cause of snow and hail, and how water contracts and becomes hard in the form of ice, and of the new shapes that the snow forms in the air . . .'[24]

If Leonardo had completed the book and had he arranged to publish it, the history of hydrology and hydraulics (or call it water resources engineering) would have undoubtedly acquired an entirely new dimension during the period of his own lifetime.

Other hydrological concepts of Leonardo
The Roman water commissioner Frontinus, some fourteen centuries before the time of Leonardo, had a vague notion that discharge through an orifice was dependent on the head of water. Leonardo's statement about it was more precise than that of Frontinus, but he too, erred somewhat. Leonardo proposed that a simple direct functional relationship existed between the head and the velocity of the discharge. He then extended this same analogy to other hydraulic problems, notably to the cases of weirs and sluice gates. He even gave an example, namely: if the head of water over a weir is increased from 2 to 3 inches, then the lower-most inch doubles its power, velocity, and discharge.[17]
Leonardo was one of the first men to study river bed configurations systematically. He was quick to realize that the velocity of water is higher at the centre of a regular channel than at the sides because of the frictional effect of the banks. He must have taken a great interest in the flow of water in rivers and channels. It is quite certain that he used models for such studies. His notebooks contain sketches of small channels in wood which were used for his experiments. He used dyes to study the motions of water, and his comments and drawings even indicate his use of glass sides on some of his flumes to facilitate better observations. One of his model studies, for example, was concerned with the nature of sedimentation behind an obstacle on a river bed. He placed sand at the bottom of a wooden channel, and placed a stone at its centre to represent the obstruction. Sandy water was used in this investigation, and it was observed that rather heavy sedimentation occurred behind the obstacle. Admittedly,

the results of most of Leonardo's model studies were qualitative rather than quantitative, as in the above example, but one must nevertheless admire him for his efforts.

Leonardo believed that the salinity of the sea was caused by various rivers which passed through salt deposits and subsequently discharged their high salt content into the ocean, and that the salinity gradually increased with the evaporation of water from the sea. He observed that sea-water contains 'more salt in our time than it has ever been at any time previously'.

The irrigation, drainage, and flood control theories of Leonardo have been discussed in considerable detail by Parsons.[25]

GERONIMO CARDANO

Geronimo Cardano (1501–1576) was a voluminous and repetitious writer. Because his works were read fairly extensively, he was quite influential during the latter half of the sixteenth century. It is said that he plagiarized freely from the manuscripts of Leonardo (which means that the Italian had at least some influence on science through such plagiarists). Cardano reduced the traditional four elements to three[26] by omitting fire. He also reduced the four qualities to two, by omitting heat from the sky and moisture from the three elements. He believed that the earth, like a sponge, was full of subterranean water,[27] and that the proportion of land greatly exceeded that of water. Water remains on the surface of the earth only because there is not enough room for it inside. This concept, he believed, is consistent with the observed fact that water always tends to flow downwards.

Cardano thought that streams start from little drains or gullies which collect condensed vapour from both above and below the ground. When rain and melted snow increase their volumes, they become rivers. He was confident that the chief source of supply of water for the rivers came from the conversion of vapour into water. Although Cardano plagiarized principally from Leonardo, he seems to have preferred Aristotle's writings on the origin of springs. The terms air and vapour were synonymous during the sixteenth century. In fact almost a century later, when discussing Cardano's concept, the eminent German geographer Bernhardus Varenius stated that there is but 'little difference between air and vapours'. Cardano had justified his conclusions with the observation that if anyone observed mountains in the early mornings, he would see a superabundance of moisture surrounding them. River stages accordingly rise in the morning, especially if they are near mountains, because of the water generated within the mountains by the coolness of the rocks and outside of them by the coolness of the nights. Later Pierre Perrault criticized[28] this opinion.

BERNARD PALISSY – THE POTTER

Bernard Palissy (figure 6) is attributed to have been born in 1499, 1510, or some time between 1514 to 1520, the most probable date being 1510.[29] He began his career making stained glass windows, but he had to do land surveying to supplement his income in order to support his wife and six children. After sixteen years of tireless experimentation, Palissy perfected a technique for making enamelled pottery which brought him fame and fortune. Later, he even applied his new art to decorating castles!

Palissy was one of the first to embrace the new Protestant religion in Saintes during Reformation and, had it not been for one of his chief patrons, Arne de Montmorency (high constable of France and the governor of Saintes) taking the case for which he was being tried, directly to the Queen Mother Catherine de Medicis, he would almost certainly have been executed. During the civil war of 1588, he was imprisoned, and died in Bastille de Bussy in 1590, an old and broken man.

Palissy loved nature, and his theories were, in general, based upon personal observations thereon. Thus, his views were often in conflict with those of the established authorities – a circumstance which did not worry him much – in fact, it added to his zeal for carrying on his own scientific investigations. His book *Discours admirables*, published in 1580, was written in French although the prevailing language of the scholars was Latin – a fact that probably explains why it never received as wide an acceptance as it deserved. Palissy himself claimed that he had no knowledge of either Latin or Greek, but Thomson[30] has suggested that his claim may have been only a pose. Duhem has claimed that the potter was a shameless plagiarist.[31]

Figure 6. Bernard Palissy (by courtesy of Librairie Honoré Champion, Paris).

Nevertheless, Palissy had a correct attitude toward experimental sciences such as paleontology, hydrology, geology, botany, agriculture, chemistry, zoology, and minerology – just to name a few of those in which he was interested. In his discourses he made use of two fictitious persons: Theory, the one who inquires, and Practice, the one who replies. Practice was the name under which he commonly distinguished his views from those of the establishment.[32] According to his sympathetic biographer Morley:

'Theory might well ask, looking back upon the whole body of doctrine taught
by the old Potter in the last years of his life,
Where have you found all this written? or tell me in what school you have been,
from which you might have learned what you are telling me.
Practice – I have no other book than the heavens and the earth, which are known
of all men, and given to all men to be known and read. Having read in the
same I have reflected on terrestrial matters, because I had not studied in astrology
to contemplate the stars.'[33]

The hydrologic cycle

It has already been shown that up to the time of Palissy very
few persons had a correct understanding of the hydrologic cycle,
but none of them had ever stated it as forcefully as the Huguenot.[34] Because Palissy was familiar with Vitruvius' work, it is
possible that he obtained his idea initially from the Roman, but
even so, one has to admire his substantiation of the theory with
actual observations of nature, carried out with sound logic by
himself.

Palissy in the guise of Practice, stated categorically that rivers and
springs cannot have any other source than rainfall, whereupon
he was called a 'great dolt' by Theory for being presumptuous
enough to attempt to contradict the most excellent philosophers
of all times. This did not worry him, for he had confidence that
his theory was correct. With great skill and logic, he refuted
the age-old theories that streams originated directly from sea-
water or from air that had been converted into water. He used
the concept of gravity effectively against the first theory by claiming
that if the theory were correct, the level of the sea would have to be
higher than that of mountains at the tops of which the rivers were
supposed to originate, that their waters would have to be saline, and
that they would have to dry up during periods of low tide. It is
true, the potter argued, that rivers do tend to dry up sometimes
but the period does not coincide with low tides; in fact, the phenomenon occurs even in summer when rainfall is lacking – a fact
which tends to support his own opinion. The maximum tidal levels
are associated with the full moons of March and July and, hence,
if the theory of his opponents was true, infinite numbers of wells and
rivers could not go dry during the months of July, August, and
September. Even assuming that the sea level was as high as the
mountains, the only way the water could ascend from the ocean

to the mountain tops under pressure would be through water-tight tunnels. Such tunnels are extremely rare and, even if they did exist in nature, water would escape through the first available hole and flood the entire countryside. Being well satisfied with the validity of his objections to the 'ocean-fed springs' concept, Theory is asked by Practice to fetch any of his Latin philosophers who could possibly present even a single contrary argument!

Palissy did not believe that air could turn into water in the vaults of caverns or cliffs, but he did concede that water could form under such circumstances by the condensation of vapour. But even then, the amounts produced by such a process would be entirely inadequate to provide enough water to sustain river-flows all over the world.

Having thus discarded the two most popular theories of his time, the potter announced his own opinion:

'rain water that falls in the winter goes up in summer, to come again in winter, and the waters and the action of the sun and dry winds, striking the land, cause great quantities of water to rise: which being gathered in the air and formed into clouds, have gone in all directions like heralds sent by God. And when the winds push these vapours the waters fall on all parts of the land, and when it pleases God that these clouds (which are nothing more than a mass of water) should dissolve, these vapours are turned into rain that falls on the ground.'[35]

This is a more precise concept than was given by most of his predecessors. But questions about what happens after rainfall and how does a river originate still persisted – and Palissy, who knew he was on the right track, was not to be put off:

'And these waters, falling on these mountains through the ground and cracks, always descend and do not stop until they find some region blocked by stones or rock very close-set and condensed. And then they rest on such a bottom and having found some channel or other opening, they flow out as fountains or brooks or rivers according to the size of the opening and receptacles; and since such a spring cannot throw itself (against nature) on the mountains, it descends into the valleys. And even though the beginnings of such springs coming from the mountains are not very large, they receive aid from all sides, to enlarge and augment them; and particularly from the lands and mountains to the right and left of these springs.'[36]

This in a nutshell, said Palissy, is the explanation for the origin of rivers and springs, and no one need seek any other reason than the present one.

Other hydrological concepts

The other hydrological contributions of Palissy, beside the firm declaration about the origin of springs and rivers, were the principles of artesian wells, recharge of wells from nearby rivers, lag in change of water levels in rivers, and forestation for the prevention of soil erosion, and plans for building 'fountains' for domestic water supply. The principle of artesian wells was explained casually[37] in a chapter on marl where the potter discussed the exploration and the stratification of soils. Wells of this type were bored in Artois long before his time, but the Huguenot provided a scientific explanation:

'I think the soil might be pierced easily by rods, and by such means one might easily discover marl, and even well-waters which might often rise above the spot at which the auger found them: and that could take place provided they came from a place higher than the bottom of the hole that you have made.'[38]

He was equally casual on the recharge of wells from waters of nearby rivers. He believed that water in wells near a river came from the river itself. This is justified because the levels of water in such wells are high during periods of high flows but when the river stages fall, so do the levels of water in wells. It conclusively proves that there are 'veins' in the earth which extend from the wells to the rivers. If the word 'vein' should be changed to bed or aquifer, this statement would express the present explanation of this phenomenon.[39] Palissy was a keen observer of nature, and drew valid conclusions from his observations. He was also a practical man, and a part of the first chapter on waters and fountains from his *Discours admirables* shows his ingenuity in applying the knowledge thus gained to the solution of problems. The problem under consideration in this chapter was how enough water could be obtained for a household situated in a sandy area. Four different cases were given consideration.

The first case had reference to a house located at the base of a mountain. He suggested that all holes and fissures in the mountain side be sealed off so as to reduce leakage to a minimum. The rain water falling within the receiving area can then be guided to discharge at the required point near the base of the mountain which can be accomplished by building walls on both sides. Large stones should be placed across the deeper channels to safeguard the house from the destructive effects of the flash floods which normally occur after heavy rains. As an extra precaution, as well as an investment, trees should be planted on the mountain side, because they will not only slow down the velocity of flow but also will be effective in preventing soil erosion. This is one of the earliest recorded descriptions of the beneficial effects forests have with respect to soil conservation. The water comes down the mountain and passes through two reservoirs, one lower than the other. The upstream reservoir is lined with sand on its uphill side to purify by filtration the water of its undesirable salts. A perforated sheet of metal at the entrance of the second reservoir assures one that only water would enter it. This water could be used for all domestic purposes.

The second case, which was just an extension of the previous one, has reference to a castle, surrounded by a moat, a reasonable distance from a mountain. The plan differs from the first in that the water from the lower reservoir is to be brought to the castle through a series of pipes.

The third case contemplates the building of a reservoir in flat terrain some distance away from a mountain. The solution to this problem is to select a plot, the size of which depends on the size of the owner's family. That plot is to be given a slope toward the house by shifting soil from one side to another (if necessary), until the difference between the ground levels at the two ends is about four feet. The surface is to be paved with rock, brick, or clay to prevent any loss of water from percolation. Holes can be left in the paving for trees, and the pavement may be covered with about a foot of soil to provide for the cultivation of vegetables. This would be a 'multi-purpose reservoir' which would provide water, fruits, wood, vegetables, as well as serve as a pasture or a very delightful place to take a walk. It would be very interesting to know if the building of such a reservoir (none has been, so far as the author could find out) would fulfill the potter's elaborate predictions. While discussing the reason why the seeds sown that way will not be submerged by water in this reservoir, Palissy gives an excellent explanation of the lag in the rise of water levels in rivers:

'rain water that falls on mountains, lands, and all places that slope towards rivers or fountains, do not get to them so very quickly. For if it were so, all fountains would go dry in summer: but because the waters that fell on the land in winter cannot flow quickly, but sink little by little until they have found the ground floored by something, and when they have found rock they follow its slope, going into the rivers. From this it follows that under these rivers there are many continual springs, and in this way, not being able to flow except little by little, all springs are fed from the end of one winter to the next.'[40]

The fourth and final case was similar to the third except that the owner is unable to afford the cost of paving. It is suggested that under such a circumstance the soil should be packed hard and covered with grassy turf, the closely intertwined roots of which would reduce the percolation losses.

One clear picture emerges from the previous descriptions of building reservoirs, that is: Bernard Palissy observed and tried to emulate nature. The basic fundamental principles underlying the four solutions are exactly the same, i.e., by selecting a gathering ground and reducing its losses due to percolation and evaporation, the maximum amounts of rain water for domestic purposes can be made available to the land owner.

GEORGIUS AGRICOLA

Georgius Agricola (Georg Bauer, 1494–1555) was quite explicit about the origin of ground-water. He, like Aristotle,[41] believed in the creation of water within the earth from the condensation of vapour. Subterranean heat primarily from burning bitumen was the cause of vapourization.[42] His explanation of the process follows:

'*Halitus* [steam] rises to the upper parts of the *canales* [openings in the earth], where the congealing cold turns into water, which by its gravity and weight again runs down to the lowest parts and increases the flow of water if there is any. If any finds its way through a *canales dilatata* the same thing happens, but it is carried a long way from its place of origin. The first phase of distillation teaches us how this water is produced, for when that which is put into the ampulla is warmed it evaporates [expirare], and this *halitus* rising into the operculum is converted by cold into water, which drips through the spout. In this way water is being continually formed underground.'[43]

Thus, ground-water primarily comes from two sources: infiltration of surface water (rainfall, river and sea), and condensation of sub-terranean steam which is generated by heating the deeply percolated surface waters. Even though both Agricola and Aristotle believed in the subterranean generation of water, the former was slightly more rational in his concept as he advocated liquefaction of vapour in contradistinction to the Stagirite who preached conversion of air into water. Even then, it was to be admitted, Agricola's suggestion of vapourization-condensation process was highly unsatisfactory.

The origin of rivers and springs was similarly attributed to the combined effect of rainfall and *halitus*. The percolation of sea-water landwards was justified by the fact that wells dug near the sea shore frequently yield saline water.

Agricola believed in the use of divining rods for locating minerals or water. This is not very surprising as even men like Robert Boyle (1626–1691), one of the founders of the Royal Society of London who lived nearly a century after Agricola, believed in the

Figure 7. Agricola's concept of the use of divining rods (from *De re metallica*).

effectiveness of the divining rod. The use of the forked twig is described in considerable detail in *De re metallica*, and figure 7 is reproduced from the book. Hoover and Hoover, translators of *De re metallica*, in a footnote state that:

'there were few indeed, down to the nineteenth century who did not believe implicitly in the effectiveness of this instrument, and while science has long since abandoned it, not a year passes but some new manifestation of its hold on the popular mind breaks out.'[44]

Ellis[45] has reviewed the use of divining rod for water witching throughout the ages.

JACQUES BESSON

Jacques Besson's life is something of a mystery, but when he wrote *L'art et science de trouver les eaux et fontaines cachées sous terre*[46] in 1569, he was a professor at Orleans. The small book (only 85 pages) was written in French, and 'the style has the same fluid character as the subject, a garrulous flow which carries the reader along without much meaning reaching the brain'.[47] The book is divided into three parts; the first part deals with the generation, place, and continuation of water above and below the ground; the second discusses the quantity, depth, and location of underground water. The last part is concerned with the treatment of unhealthy and harmful waters and with methods of conveying water.

Besson believed that at the beginning of the world there were only two elements, water and earth, and that the other two elements, air and fire, were later drawn by attenuation and resolution of a great mass of water which inundated everything. It was, undoubtedly, a biblical outlook because it subscribed to the concept that water covered the entire earth prior to its recession to the depth of the sea and underground.

Besson's main contribution to hydrology was his clear and correct explanation of the hydrologic cycle. He believed that water was evaporated by the heat of the sun, and that it then came down as rainfall. Precipitation is enough to sustain the flows of rivers and springs; evaporation and precipitation are equal in amount. The rivers carry their waters to the sea. His explanation for the salinity of the sea was rather original: it was saline right from the beginning of the creation and hence needed no explanation at all!

He rejected the theory that springs originated in the mountain tops, and stated categorically that springs on the mountain sides are results of precipitation, as can be ascertained by the fact that their flows decrease or increase with the scarcity or abundance of rainfall. There is considerable controversy over the originality of Besson's work.[47-49] Besson was a contemporary of Bernard Palissy and there are remarkable similarities in the writings of these two Frenchmen, especially with regard to the origin of springs and fountains. The dates of publication of their books are as follows:

1563: Bernard Palissy: Recepte véritable,

1569: Jacques Besson: L'art et science de trouver les eaux et fontaines cachées sous terre,

1580: Bernard Palissy: Discours admirables.

It is significant that neither Besson nor Palissy ever mentioned each other's name in their works. It is possible that Palissy first presented his theory about springs in 1563 and that it was then taken up and elaborated upon some six years later by Besson; then in 1580 Palissy finally returned the compliment. It is very difficult to believe that each of them advanced the same idea without being aware of the other's concept, and it is quite likely that each may have appropriated a large part of the other's views. On the other hand, it is quite possible, especially since both of them were attached to the French court, that Besson and Palissy knew each other. Both were interested in this same subject, and they might well have discussed it together on several occasions. And when the time came to put their thoughts on papers, each may have thought that he was the one who originated the idea. This offers, no doubt, a compromise solution to this problem, but in the absence of better evidence, it would be futile to charge either one as being a plagiarist.

FLOOD STUDY OF GIOVAN FONTANA

The architect Giovan Fontana da Meli carried out investigations of the rise of the Tiber at Rome and the consequent flooding during the Christmas of 1598. The treatise, which was dedicated to Pope Clement VIII, was first published in 1599 and was later reprinted in 1640[50] at the recommendation of Benedetto Castelli. There were numerous controversies over the flooding of the Tiber which, according to Fontana, resulted from man's ignorance of the places and sites where the many different rivers and fosses (canals, ditches, or trenches) entered the Tiber, as well as his ignorance of the effects produced by heavy rainfall and by the sirocco winds, both of which had prevailed for many months prior to the occurrence of the flood. The measurements, on which the study was based, were carried out by Fontana, his nephew, and some 'skilled' men. They systematically traced all of the rivers, torrents, and fosses which flow into the Tiber from Perugia and Orvieto down as far as Rome. Their method was to determine from the depths and the widths of flows in the water courses under normal as well as flood conditions, their wetted cross-sectional areas. Fontana followed the same procedure that the Roman water commissioner, Frontinus had used, to determine the discharge through open channels, that is the '$Q = A$' concept (discharge is equal to the wetted cross-sectional area). Thus it can be concluded that after the passage of nearly 1500 years, no improvement had taken place in the technique of measurement of flow in open channels. A sample calculation, and the conclusions of Fontana's investigation are presented below:

'The ordinary running water of the bed of the Chiane River is 72 palms wide and has a depth of water of 8 palms. The Chiane river in the flood which happened at Christmas was 152 palms wide and rose in the flood above today's running water 18 palms $^1/_2$. Thus discharge during the flood was $152 \times 18^1/_2 = 2812$ square palms (28 ells 12 palms).'

The calculation is divided into 5 parts: water brought in by (1) the river Chiane, (2) the river Paglia, (3) the river Tiber in the place of the Castello di Corbara nearly a mile upstream from the mouth of the Paglia river, (4) the rivers and streams entering the Tiber, from the Corvara to Rome from east and west sides, and, (5) the river Nera and its tributaries. The final result was as follows:

Total, Paglia river	73 ells 26 palms
Total, Chiane river	28 ells 12 palms
Total, Tiber from Castel della Corvara up to where it begins	135 ells 90 palms
Total, water brought into Tiber rivers and fosses from Corvara to Rome	219 ells 61 palms

Total, river Nera with all its tributaries	42 ells 50 palms
Grand total	500 ells 9 palms

From his investigations, Fontana concluded that the area above Rome contributed to the river Tiber 500 square ells and 9 palms of water more than it normally carried. His recommendation was: 'to eliminate the flood at Rome, it would be necessary to provide two more river beds as ample as the one there is today, at least'.[50] The solution proposed, in simple words, was channel improvement. The systematic measurement and analysis of flood flow, though conducted under an erroneous concept, undoubtedly turned out to be a praiseworthy effort. No wonder Benedetto Castelli, who in 1628 discussed the full effect of velocity on discharge,[51] recommended that the report of Fontana's investigation be reprinted.

CONCLUSION

The end of the sixteenth century marks the closing of a distinct era in the history of hydrology. Up to that time it was believed that discharge in open channels or pipes was equal to the wetted cross-sectional area. Had the hydrologist of the era expressed this concept mathematically, it would have been '$Q = A$'. It was the same concept that was used many centuries earlier by Frontinus and, it indicates that the methodology for computation of discharge remained the same for more than fifteen centuries. It is true that both Hero and Leonardo had a correct understanding of the phenomenon but, unfortunately, their works, in this particular field, seem to have gone completely unnoticed.

The contributions of Leonardo and Palissy, as far as hydrology is concerned, have one common aspect, that is they did not affect the immediate development of the science. Leonardo's notes were generally not available, and even though Palissy's works were printed, they were in French and, hence probably, did not attract the attention they deserved. According to Pierre Duhem,[52] even the most novel and audacious intuitions of Leonardo had been suggested and guided by the medieval science, and he cannot be regarded as an isolated example who was neither influenced by the past nor had any influence on the future. A similar stand has been taken recently by Bertrand Gille[3] who claimed that Leonardo's works were largely dependent on those of Francesco di Giorgio and other engineers. Duhem further accused[53] Palissy of being guilty of extensive plagiarism. The originalities of Leonardo and Palissy may be debatable, but it has to be admitted that they were the most outstanding figures of their times in the field of hydrology.

Initially, Leonardo accepted the classical and medieval concept that heat provides energy for elevation, which led him into futile speculations about the origin of rivers and springs – but, he did achieve some results. He was especially lucid in his explanation of continuity, in streams, and his experimental determination of the distribution of velocity in open channels was certainly noteworthy. Palissy can be commended for his views on the origin of springs and rivers, artesian wells, and water resources engineering in general.[54] Both of them opted for experimental methods, but both were equally adept in passing some 'old wives' tales. Undoubtedly their true positions will be somewhere below the statements of hero-worshippers but higher than what the skeptics are willing to admit.

The chapter can best be concluded with a quotation from Leonardo:

'Wisdom is the daughter of experience ... No human enquiry is worthy of the name of science unless it comes through mathematical proofs. And if you say that the sciences which begin and end in the mind possess truth, this is not to be conceded, but denied for many reasons. First because in such mental discourses there enters no *esperienza** without which nothing by itself reaches certitude.'

REFERENCES

1. HALL, A. R. and M. B. HALL, A brief history of science. New York, New American Library of World Literature Inc. (1964) p. 73.
2. THOMSON, S. H., Europe in Renaissance and Reformation. New York, Harcourt, Brace and World Inc. (1963) pp. 348–349.
3. B. GILLE, The Renaissance engineers. London, Lund Humphries (1966).
4. HART, I. B., The world of Leonardo da Vinci. London, Macdonald & Co. Ltd. (1961) p. 19.

* Wavers in meaning between experience and experiment.

5. ADAMS, F. D., The origin of springs and rivers. Fennia *50* (1928) 3–16.
6. ADAMS, F. D., The birth and development of geological sciences. Baltimore, Williams & Wilkins (1938). Reprinted by Dover Publications Inc., New York (1954) pp. 426–460.
7. MEINZER, O. E., The history and development of ground-water hydrology. Journal of the Washington Academy of Sciences *24* (1934) 6–32.
8. MEINZER, O. E., Hydrology. Physics of the earth, vol. IX. New York, McGraw-Hill (1942) pp. 8–30.
9. ADAMS, F. D., Ref. 6, pp. 445–446.
10. DE LORENZO, G., Leonardo da Vinci e la geologia. Bologna (1920) pp. 109–111.
11. KRYNINE, P. D., On the antiquity of 'sedimentation' and hydrology (with some moral conclusions). Bulletin of the Geographical Society of America *71* (1960) 1721–1726.
12. LEONARDO DA VINCI, The literary works of Leonardo da Vinci, edited by J. P. Richter, vol. 2. Oxford, University Press (1939) p. 144.
13. *Ibid.*, p. 159.
14. *Ibid.*, p. 149.
15. LEONARDO DA VINCI, The notebooks of Leonardo da Vinci, edited by E. MacCurdy, vol. 1. London, Jonathan Cape (1956) pp. 654–655.
16. BISWAS, ASIT K., Leonardo da Vinci and the hydrologic cycle. Civil Engineering, ASCE *35* (1965) 73.
17. LEONARDO DA VINCI, Manuscript volume, edited by L. M. Arconati. Vatican Library (1643). Later published as: De moto e misura dell'acqua, with E. Carusi and A. Favaro as coeditors, Publicazioni dello Istituto di Studii Vinciani, Nuova Serie, vol. 1. Bologna (1923).
18. LEONARDO DA VINCI, Del moto e misura dell'acqua, edited by Francesco Cardinali. In: Raccolta d'autori Italiani che trattano del moto dell'acque. Bologna (1826) pp. 273–450.
19. LEONARDO DA VINCI, Ref. 15, p. 330.
20. INCE, S., A history of hydraulics to the end of the eighteenth century. Dissertation. Iowa City, State University of Iowa (1952) pp. 47–48.
21. LEONARDO DA VINCI, Ref. 15, vol. 2, p. 18; ref. 15, vol. 2, p. 70.
22. Quoted by H. ROUSE and S. INCE, History of hydraulics. Iowa Institute of Hydraulic Research, State University of Iowa (1957) p. 49.
23. MARINONI, A., Tempo armonico o musicale in Leonardo da Vinci. Lingua nostra, vol. 16, fasc. 2, June, 1955.
24. LEONARDO DA VINCI, Ref. 12, p. 152.
25. PARSONS, W. B., Engineers and engineering in the Renaissance. Baltimore, Williams & Wilkins Co (1939).
26. CARDANO, J., The first book of Jerome Cardan's De subtilitate, translated by M. M. Cass. Williamsport, Pa., The Bayard Press (1934) p. 15.
27. CARDANO, J., De rerum varietate libri XVII, I, 1–2, X, 49. Basileae, H. Petri (1557).

28. PERRAULT, P., De l'origine des fontaines. Paris, Pierre Le Petit (1678) pp. 41–49.

29. LEROUX, D., La vie de Bernard Palissy. Paris, Librairie Ancienne Honoré Champion (1927) p. 13.

30. THOMPSON, H. R., The geographical and geological observations of Bernard Palissy the Potter. Annals of Science *10* (1954) 149–165.

31. Quoted by L. THORNDIKE, A history of magic and experimental science, vol. V. New York, Columbia University Press (1941) pp. 596–599.

32. BISWAS, ASIT K., The hydrologic cycle. Civil Engineering, ASCE *34* (1965) 70–74.

33. MORLEY, H., The life of Bernard Palissy, of Saintes, 2nd ed. London, Chapman & Hall (1855) pp. 471–472.

34. PALISSY, B., Discours admirables. Paris, Martin Le Jeune (1580).

35. PALISSY, B., The admirable discourses, translated by A. la Rocque. Urbana, University of Illinois Press (1957) p. 53.

36. *Ibid.*, p. 56.

37. *Ibid.*, p. 218.

38. MORLEY, H., *op. cit.*, p. 471.

39. PALISSY, B., Ref. 35, p. 34.

40. *Ibid.*, pp. 67–68.

41. ADAMS, F. D., Ref. 6, p. 445.

42. AGRICOLA, G., De ortu et causis subterraneorum lib. V. Basileae, H. Frobenium et N. Episcopium (1546).

43. AGRICOLA, G., De re metallica, translated by H. C. Hoover and L. H. Hoover. New York, Dover Publications Inc. (1950) pp. 46–48.

44. *Ibid.*, p. 38.

45. ELLIS, A. J., The divining rod – a history of water witching. Water-Supply Paper 416, Geological Survey, U.S. Government Printing Office, Washington D.C. (1917).

46. BESSON, J., L'art et science de trouver les eaux et fontaines cachées sous terre. Orléans, P. Trepperel (1569).

47. THORNDIKE, L., *op. cit.*, pp. 596–599.

48. THOMPSON, H. R., *op. cit.*, pp. 150–165, no. 2.

49. PALISSY, B., Ref. 35, p. 13.

50. FONTANA, G., Dell' accrescimento che hanno fatto li fiumi, torrenti, e fossi che hanno causato l'inondation à Roma il natale, reprinted at recommendation of Sig. Domenico Castelli. Gioiosi, Appresso Antonia Maria (1640).

51. CASTELLI, B., Della misura dell'acque correnti. Roma, Nella Stamparia Camerale (1628).

52. DUHEM, P., Etudes sur Léonard de Vinci, vol. 1. Paris, A. Hermann (1906) pp. 182–183.

53. DUHEM, P., *op. cit.*, 3 vols. (1906), (1909), (1913).

54. PALISSY, B., Resources: a treatise on waters and springs, translated by E. E. Willett. Brighton, D. O'Connor (1876).

9

The seventeenth century

INTRODUCTION

Very often historians call the seventeenth century 'the cradle of modern science', because it started with so little knowledge and ended with so much. It made impressive and significant contributions such as Galileo's mechanics, Kepler and Newton's astronomy, Harvey's blood circulation, Descartes' geometry, Van Leeuwenhoek and Hooke's microscopy, and last but not least, Perrault, Mariotte and Halley's experimental investigations which produced a concept of the hydrologic cycle. This was the period that saw the downfall of Aristotle, and the remoulding of man's mind, by replacing the teleological aspects of the previous centuries with experimental philosophy. John Dryden (1631–1700) in his poem *Longest tyranny*, written in 1663, the year of his election to the Royal Society, said:

'The longest Tyranny that ever sway'd
Was that wherein our Ancestors betray'd
Their free-born *Reason* to the *Stagirite*
And made his *Torch* their universal *Night*.
So Truth, while onely *One* supplies the State,
Grew scarce, and dear, and yet sophisticate*;
Until 'twas bought, like Empirique Wares, or charms,
Hard words sealed up with Aristotle's Armes**.'

---

* In the seventeenth century English, sophisticate meant adulterated.
** The last two lines mean that obscure and dubious works were made significant by just being attributed to the authority of Aristotle, the great master.

Aristotle thought that 'all men by nature desire to know', but it took Johannes Kepler and some two thousand years to state that 'to measure is to know'. According to Francis Bacon (1561–1626):

'There are two ways, and can only be two of seeking and finding truth. The one, from sense and reason, takes a flight to the most general axioms, and from these principles and their truth, settled once for all, invents and judges of all intermediate axioms. The other method collects axioms from sense and particulars, ascending continuously and by degrees so that in the end it arrives at the most general axioms. This latter is the only true one, but never hitherto tried.'[1]

The various developments in the field of hydrology during the seventeenth century are discussed herein, except those of Pierre Perrault, Edmé Mariotte, and Edmond Halley. They will be treated separately in the next chapter.

GALILEO, KEPLER AND DESCARTES

Galileo Galilei (1564–1642) started his life with the traditional scholastic teaching of Aristotelean theories, but he soon abandoned them. In 1585, he started to conduct experimental investigations with some of Aristotle's doctrines and, by 1590, proved that some of them were wrong. Galileo did not contribute much to the development of hydrology directly, but his attitude towards science had a profound effect on all branches of knowledge, and hydrology was no exception. Both Benedetto Castelli and Evangelista Torricelli, his students, were considerably influenced by this Italian polymath.

The German, Johannes Kepler (1571–1630), a contemporary of Galileo and a man of strong mystical tendencies, was the real founder of scientific heliocentrism.[2] His work was responsible, to a certain extent, for the downfall of Ptolemy; just as Galileo's experiments helped destroy the reputation of Aristotle. His idea of the universe was somewhat Platonic and Pythagorean. During the early seventeenth century, the concept that the earth is a living being or at least it functions like one, gained some supporters, including Kepler. Adherents of this principle could be found even as late as the nineteenth century. The theory can be seen in Kepler's *Harmonices mundi*, published in 1619, which dealt with his third and the last great law of the planetary motion. According to it:

'The globe contains a circulating vital fluid. A process of assimilation goes on in it as well as in animated bodies. Every particle of it is alive. It possesses instinct and volition even to the most elementary of its molecules, which attract and repel each other according to sympathies and antipathies. Each kind of mineral substance is capable of converting immense masses of matter into its own peculiar nature, as we convert our aliment into flesh and blood. The mountains are the respiratory organs of the globe, and the schists its organ of secretion. By the latter it decomposes the waters of the sea, in order to produce volcanic eruptions. The veins in strata are the caries or abscesses of the mineral kingdom, and the metals are products of rottenness and disease to which it is owing that almost all of them have so bad a smell.'[3]

The earth drinks in sea-water which undergoes the process of digestion and assimilation, and the end product of these physiological processes is discharged through springs.[4, 5]

René Descartes (figure 1), son of a lawyer, was born at La Haye in Touraine, in 1596. He was educated at the Jesuit College at La Fleche, in Anjou, and later at the University of Poitiers from where he graduated in Law, in 1616. He travelled extensively in Europe and came to know people like Mersenne and Picot. In 1628, he discarded his nomadic life in favour of working to establish a new philosophy, free from the influences of ancient doctrines. He went to Holland to carry out his studies in tranquility (because the country was peaceful) under the Stadholder, Prince Frederic Henri, who had a reputation for encouraging religious freedom, which was rather unusual for that time. The book *Discours de la méthode* (Discourse on method) with three accompanying essays, *La dioptrique* (Dioptrics), *Les météores* (Meteors), and *La géométrie* (Geometry) was published in 1637. He died in Stockholm, in 1650. Descartes, like Herodotus and Bacon, considered all knowledge to be within his province, but mathematics was the particular subject in which he found real satisfaction. The major work, so far as hydrology is concerned, is his essay on meteors. He attempted to attribute natural causes to natural phenomena, and even though his reasoning was not always correct, he at least separated them from the realm of magic and occult.

Les météores is probably the most original work on the subject since Aristotle's *Meteorologica* and, in it, Descartes tried to explain various meteorological phenomena on a 'scientific' basis. It is true that the work does not reach the same high standards as his writ-

Figure 1. René Descartes (by courtesy of the Royal Society of London).

ings on mathematics or even optics but, nevertheless, it is quite a significant piece of writing. He believed that one needs to know about the structures of air, water, and the earth, as well as the bodies on it, to understand various natural phenomena. They are composed of infinite numbers of little parts of different sizes and shapes which, as they are not compactly joined, have gaps in between them, filled with a fine 'subtle matter'. The component parts of water are long and smooth, and can be easily separated as they never get hooked together. In contrast to water, air, and other bodies, have very irregular components. If there are many gaps around the constituents that are filled with the subtle matter, the material becomes very rare and light, like air or oils. Even the formation of ice and expansion of water while freezing were explained by the presence of the subtle matter. Descartes, however, did believe some 'old wives' tales:

'And we see by experience that water which has been kept on a fire for some time freezes more quickly than otherwise; the reason being that those of its parts which can be most easily folded and bent are driven off during the heating, leaving only those which are rigid.'[6]

Had he conducted a simple experiment, he would have undoubtedly found out the truth about this medieval concept, but it indicates that Descartes was not entirely free from relying on *a priori* conclusions.

He explained the process of evaporation by the presence of the subtle matters which are agitated by the sun or some other cause. The subtle matters, in turn, agitate the bodies which envelop them. The smallest parts become detached and rise up – not because of their natural tendency or even the attraction of the sun but due to their motion that cannot continue in any other direction. The land and sea breezes are caused by the expansion of vapours, as they have a tendency to move to those regions where they can find more space. Clouds and fogs are formed when the expanded vapours condense and become more compact. As the vapour particles get chilled they lose their mobility – thus forming water droplets or ice crystals. Since for a given volume a sphere has the least surface-area, water droplets generally become spherical. Their shapes might become changed because of the influence of other forces, i.e., air resistance during rainfall. A single droplet remains suspended in the air but when a few of them combine, their total weight becomes too great, and they consequently fall either as rain or dew.

The formation of snow or hail was explained in a different manner. Snowflakes, which are frozen and expanded water, are light and do not generally fall to the earth. But sometimes, because of certain

conditions like the expansion of air above the clouds, they do come
down. If they are melted completely by warm air while falling,
rainfall occurs; but if they remain unmelted snowfall; while if,
after having melted, they meet cold wind, hail is produced.

Descartes' concept on the origin of springs and rivers was curious,
to say the least, but his theory was the dominant one for nearly two
centuries. He maintained that the sea-water diffuses through a
series of subterranean channels, in various directions, until it
reaches large caverns at the base of a mountain. There the water
evaporates due to the heat of the earth's interior, and the salt is left
behind (because it is too 'gross' and heavy). The vapour is subsequently condensed by the low temperature at the top of the vaults,
and the water produced there emerges as streamflow.[7-9] Curious-
ly, nowhere does he mention what happens to the enormous deposit
of salt that would have accumulated if his theory had been true.
Probably *Les météores'* greatest claim to fame is the section on
rainbows which is exhaustive and extremely well-written, but it
does not have any direct connection to the science of hydrology.

CASTELLI ON DISCHARGE CALCULATIONS

Benedetto Castelli (1577–1644; figure 2), born at Brescia, was
entered into the Benedictine monastery of Montecassino at an
early age. The Montecassino, as it may be recalled, had already
ascertained a unique place in the history of hydrology by preserv-
ing the irreplaceable work of Frontinus for posterity. Castelli
was a student of Galileo and later became one of his most trusted
friends. When Galileo's theories on hydrostatics were attacked, he
vigorously defended them.

Castelli, a mathematician to the Pope Urban VIII, taught mathematics at the universities of Rome and Pisa. In the preface of
the book *Della misura dell'acque correnti*,[10] published in 1628, Castelli
stated that he was ordered by the Pope to apply his thoughts to
the motion of water in rivers – a subject which is difficult, most
important, and very little considered by others. His teacher
Galileo once said that 'I can learn more of the movement of Jupiter's satellites than I can of the flow of a stream of water',[11] and

Figure 2. Benedetto Castelli.

no wonder, when Castelli's work was published, it was proclaimed
by his teacher to be a 'golden book'.

The main contribution of the Benedictine monk to hydrology
is his clear explanation of the relationship between velocity and
discharge of flow, a concept put forward by both Hero and Leonardo, but in those instances, had gone completely unnoticed.
The book[10] restated the principle of continuity most convincingly
and, for that reason, he is often called the father of the Italian school
of hydraulics. The principle, as suggested by Castelli, was that:

'... in divers parts of the same river or current of running water, there doth
always passeth equal quantity of water in equal time and it being also true, that
in divers parts the same river may have various different velocity; it follows of
necessary consequence, that where the river hath less velocity, it shall be of
greater measure, and in those parts in which it hath greater velocity, it shall be
of less measure; and in sum, the velocity of several parts of the said river, shall
have eternally reciprocal and like proportion with their measures.'[12]

From the above theory, he derived five axioms. They are:

Axiom I: 'Sections equal, and equally swift, discharge equal quantities of
water in equal times.'

Axiom II: 'Sections equally swift, and that discharge equal quantity of water,
in equal time, shall be equal.'

Axiom III: 'Sections equal, and that discharge equal quantities of water in
equal times, shall be equally swift.'

Axiom IV: 'When sections are unequal, but equally swift, the quantity of the
water that passeth through the first section, shall have the same proportion
to the quantity that passeth through the second section. Which is manifest,
because the velocity being the same, the difference of the water that passeth
shall be according to the difference of the sections.'

Axiom V: 'If the sections shall be equal, and of unequal velocity, the quantity
of the water that passeth through the first, shall have the same proportion
to that which passeth through the second, that the velocity of the first section,
shall have to the velocity of the second section. Which also is manifest, because
the sections being equal, the difference of the water which passeth, dependeth
on the velocity.'[13]

He vigorously criticized[14] Fontana's work[15] on floods because of
the erroneous '$Q = A$' concept. He pointed out that since Fontana

had completely neglected the velocity of flow, the recommendation of widening the river channel could not be correct.

The originality of Castelli is extremely difficult to ascertain as it is quite possible that he was familiar with the Vatican Compilation of Leonardo's notes and, thus, may have been deeply influenced by it.[16] It is certainly remarkable, as will be seen in the next subsection, that he made the same mistake as Leonardo on the efflux problem, but that really does not prove anything. Lombardini,[17] for example, considered it to be 'an exceedingly unpleasant task' to expose Castelli as a plagiarist. He concluded that:

'. . . the hydraulic science was without a doubt created by Leonardo, but that in truth, the engineers of Lombardy were unable to consult and profit by his writings until about 1570, because the various propositions had become so scattered, and were expounded in such a way that it was difficult to read them. Castelli must have become acquainted with them 60 or 70 years later, although he maintained that he had deduced by experiments, propositions which in form and terminology, appear to have been obtained directly from the writings of Leonardo.'[18]

Poggendorff, on the other hand, considered[19] Castelli's book to be the first containing correct principles of flow of water in rivers and canals. Since it is possible that Castelli was wholly independent of Leonardo's influence, he should not, in all fairness, be called a plagiarist – at least not until some convincing evidence has been found. Be that as it may, no one would deny that he has done a great service to the science of hydrology by the much-needed correction in the discharge formula.

Castelli also made the first rain gauge in Europe, but more about that will be discussed in chapter 11.

THE EFFLUX PRINCIPLE

It was known, at least vaguely, from the time of Frontinus that the quantity of water flowing outward through an orifice at the bottom of a vessel was related to the head of water which existed above such an orifice. The problem was still not completely solved until the beginning of the seventeenth century. Galileo, in course of his experiments, for example used to measure time by the relative weights of water discharged from a large bowl of water with an

Figure 3. Athanasius Kircher.

orifice in its bottom. He and his son had both tried to build a water clock, but because of the error in that concept, their ventures failed to succeed. Probably Castelli's interest in the efflux problem was aroused by his master's makeshift water clock but he did not make much progress thereon. It was his belief that velocity depended directly on the height of water. In all fairness, however, it should be pointed out that he was not happy with this belief. The problem was first solved by Evangelista Torricelli (1608–1647) who is sometimes credited with the invention of the barometer.[20] It was Torricelli, who first stated that the velocity of efflux is dependent on the square root of the head. The deterministic equation $v = \sqrt{2gh}$, however, was not actually written until about 1738. The value of g, which was determined by Christian Huygens (1629–1695) in 1673, first had to be evaluated, and as a matter of fact, that equation was not expressed until the two Bernoullis did so in 1748. Among others who worked theoretically and experimentally, on the development of the velocity formula were Maggiotti, MacLaurin, Poleni, Newton, Guglielmini, Grandi, Mariotte and Michelotti. Frisi has admirably summed up[21, 22] the development of the efflux principle to about the middle of the eighteenth century.

ATHANASIUS KIRCHER

The German Jesuit, Athanasius Kircher (1602–1680; figure 3), was a professor at Würtzburg, who wrote a number of bulky treatises on many subjects. The book *Magnes, sive de arte magnetica* was published in Rome in 1641, and his most important work on the subterranean world,[23] *Mundus subterraneus*, came out in 1664. Both the books indicate that Kircher had an extremely fertile imagination, so much so that the book on magnetism had a separate section on the magnetism of love. The other book deals with everything within the earth, including the origin of springs and rivers. *Mundus subterraneus* became a standard geology textbook in the seventeenth century.[24] On the subject of the origin of rivers and springs Kircher was unable to free himself completely from the centuries-old domination of the Church. He started with the fundamental concept from *Ecclesiastes*, namely that rivers receive their supply of water from the sea. He considered Aristotle's concept of transformation of air into water to be rather ludicrous. He, however, used part of the Stagirite's idea for promoting his own concept. He thought that there are many great hydrophylacia (caverns containing water) within the major mountain ranges of the world, and that they were formed by God in his great wisdom during the creation of the world. Rivers flow out of these caverns in various parts of the world in order that man can use them either for irrigation or navigation. Since the quantity of water in a hydrophylacium is not limitless (as stated in Plato's Tartarus), he had to seek another explanation for their sources of supply. He soon found it in a verse from *Ecclesiastes*. There were two major problems or rather handicaps for the Jesuit to surmount, and one must give him credit for identifying them, even though his explanations were wrong. This is much better than some of his more illustrious predecessors who did not even recognize them. The difficulties lie in the nature of the connections between the hydrophylacia and the sea, and in the problem of raising sea-water to a higher level than it was originally. Kircher believed that such difficulties could be overcome, saying that 'there is no man with mind so dull, as not to be ready to follow on hands and feet, as the saying is, my trains of thought'.[23] Solving the first problem was comparatively easy, as he visualized sea-water passing to a hydrophylacium through openings in the ocean-floor. Figure 4 is an idealized sketch of a mountainous area near the sea. The whirlpools in the figure indicate the locations of openings in the sea-bed through which water is carried by subterranean channels to the caverns of mountains from where the rivers originate. The water is returned to the sea by the rivers, thus completing his version of the hydrologic cycle. In the diagram, the subterranean channels appear in a darker shade whereas the rivers are shown in a lighter shade. Figure 5 represents a section of a mountain with a cavern being supplied through subterranean channels with water from the sea.

Various theoretical processes for getting water to flow uphill (from the sea to mountain tops) by mechanical methods were explained with elaborate diagrams. A brief discussion of three of those processes is presented here. The first contemplated the use of a pair of double bellows powered by water-wheels to raise the water, and the second, a U-tube filled with water, which had one

Figure 4. Kircher's explanation of the origin of rivers and springs.

limb shorter than the other. The opening of the shorter limb was to be covered with a flexible diaphragm. When pressed, it would raise the water in the other arm higher than it was originally. The third contemplated water being raised by the creation of a vacuum. Kircher experimented with all those methods to assure himself that they would work. Having done that, he searched for an analogous process in nature, and finally arrived at an answer with which he was satisfied. It was that tides, which are created in the sea by the attraction of the moon, would cause one mass of water to acquire a higher level than the other. This condition would produce the necessary pressure, as could be demonstrated by his

Figure 5. Kircher's view of a cavern being supplied with sea-water.

U-tube experiment. This provided the justification he needed to claim that sea-water is forced through the openings in the bed, and that it flows under pressure to the mountain tops. High winds contributed a share of the pressure on the ocean surface, and they helped to force water through the subterranean channels created by God in his divine wisdom (just like the veins and arteries in the body of man, the microcosmos, to accomplish the purposes for which they were created).

A second method by which sea-water could be raised to higher than its original level was claimed to be produced by the action of fire. Kircher believed that the underground world is:

'a well fram'd house with distinct rooms, cellars, and store-houses, by great art and wisdom fitted together; and not as many think, a confused and jumbled heap or chaos of things, as it were, of stones, bricks, wood, and other materials, as the rubbish of a decayed house, or a house not yet made.'[25]

The caverns are occupied by either fire or water, and all the fire-

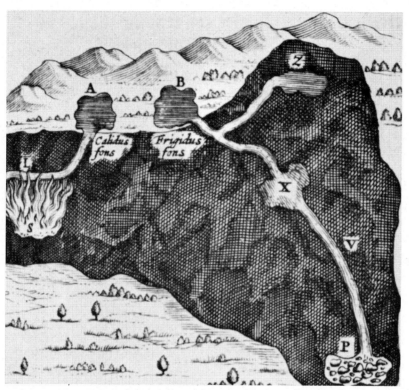

Figure 6. Origin of hot and cold springs according to Kircher.

filled ones have a direct communication with the central fire. If the fire is near the surface, it could break out as a volcano, but if it is located deeply within the earth, it heats up the water in a nearby cavern, assuming, of course, that such a cavern exists. The water vapourizes, comes towards the surface, recondenses, and thus generates hot springs. Figure 6 shows a hot and a cold spring with origins close to one another. The explanation was very simple: the hot spring A is created by the passage of the subterranean channel L over the fire-filled cavern S, whereas the spring B is cold as there is no fire nearby. If, after being heated as in the case of spring A, water has to travel a long distance before it comes to the surface, cooling takes place and a cold spring is produced.

Formation of mineral springs is shown in figure 7. Issuing from a

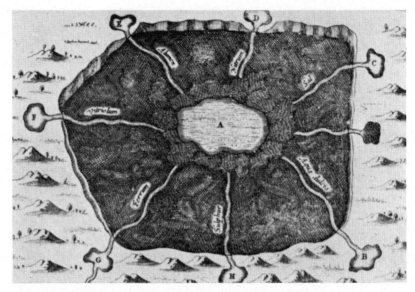

Figure 7. Kircher's explanation of formation of mineral springs.

common hydrophylacium A, the water qualities of the springs change according to the mineral substances they encounter enroute. For example, spring H passes through sulphur-bearing rocks and, hence, it is sulphurous, whereas the water from spring B is pure because it does not pass through any soluble materials.

It is evident that Kircher was deeply interested in groundwater and origin of springs and rivers but, unfortunately, his basic concepts were not correct.

SCHOTT AND BECHER

An abridged version of Kircher's views was first published in 1656.[26] Three years later, another Jesuit, Gaspard Schott (1608–1666), treated the subject of rivers and springs rather exhaustively. He disagreed with Kircher, and interpreted the verse from *Ecclesiastes* to suit his own purposes:

'We are of the opinion that some springs and rivers have their origin from subterranean air and vapours which have been condensed into water. Others from rain and snow which have soaked into the earth, the greatest number and the most important rivers, however, from sea water rising through subterranean passages and issuing as springs which flow continuously. And so the sea is not the only source, at least it does not distribute its water through underground passages to all these springs and rivers. But this statement would seem to run contrary to the clear teaching of Holy Writ found in Ecclesiastes, chapter 1 and verse 7, *All rivers run to the sea; yet the sea is not fully unto the place whence the rivers come, thither they return again.* The real meaning of these words however seems to be: All rivers run into the sea, from the place out of which they come, to it they flow back again. Consequently those which enter the sea have issued from the sea, and those which have issued from the sea return to it and enter it that they may flow out of it again. But all enter it and all return to it, therefore all have issued from it. But it does not follow that some, as we believe, have not come out of the sea by another road than that just mentioned. I am, therefore firmly of the opinion and again repeat, all rivers do not issue from the sea – at least all do not make their exit directly out of the ocean into the depths of the earth and from there rise through subterranean channels to their fountain heads. This is held to be true not only by recent authorities as, 'Conimbricensium' Fromondus, Cabeus, Cornelius, Magnanus, as seen in their words above referred to, but also by Albertus Magnus, (Duns) Scotus and a multitude of others who believed it to be consonant with the teaching of Scripture.'[27, 28]

Johann Joachim Becher (1635–1682) was one of the most well-known alchemists of the seventeenth century. He believed that there was a large vaulted space in the centre of the earth which was extremely hot due to the ignition of sulphurous and bituminous materials. The water from the sea comes to the fiery centre through various fissures, and then evaporates. The vapour rises through the earth's crust to the tops of mountains where it condenses and gives rise to springs and rivers.[29] Thus, Becher's theory was somewhat similar to that of Descartes: the only difference is that the alchemist believed that water evaporated at the centre of the earth whereas Descartes believed that evaporation took place in caverns located at the base of mountains. These basic concepts, quite common during the sixteenth and the seventeenth centuries, all show a similarity to the working of an alembic (figure 8). Before the idea had been conceived, it had been commonly believed that salt water, during its travel, passed through fissures that were so narrow that only pure water could pass through them and that the salt was left behind. When it was realized that salt can be separated from sea-water only by distillation (and not by filtration), the new alembic theory began to receive popular acceptance.

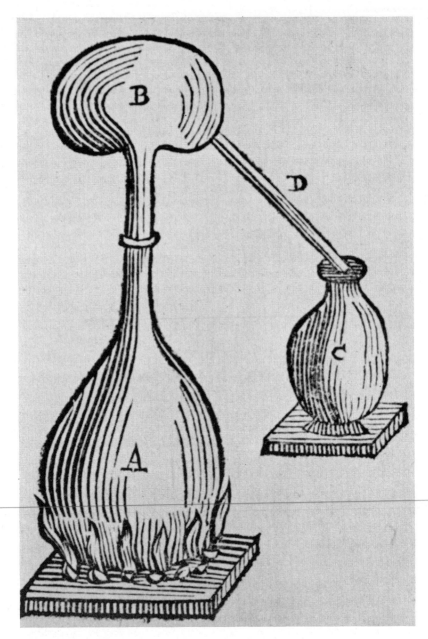

Figure 8. An alembic.

BERNHARDUS VARENIUS

Bernhardus Varenius (1622–1650), born in Hanover, was trained in medicine at Königsberg and Leyden. He settled in Amsterdam and wrote two books, *Descripto regni Japonise* and *Geographia generalis*, which were published in 1649 and 1650 respectively. He was probably the most famous geographer of his time, and his thoughts influenced geography for more than a century. He died in Leyden, in 1650, at the early age of twenty-eight.

The concepts of Varenius on the origin of springs and rivers were a classic case of compromise.[30] He believed it most reasonable to assume that the sea-water hypothesis was correct, but did not exclude the two other factors – precipitation and the conversion of air into water. He advanced two reasons for opting in favour of the sea-water hypothesis. First, rivers bring in a vast amount of water to the sea, and since the sea does not increase in volume, sea-water has to be 'refunded out of the sea into the earth and carried to the heads of rivers'.[31] This can be substantiated by actual observations, namely, the closer a river comes to the sea, the more brackish is its water. Some salt springs are actually supplied with water through direct subterranean conduits from the sea. Second, the sea-water conversion theory alone can explain the presence of water at great depths, as in mines, where it cannot come from rainfall or from the condensation of air. The latter was, in fact, just a restatement of Seneca's theory.

However, he still needed to explain how the sea-water reached the heads of rivers and managed to lose its salinity in that process. This he did by saying:

> 'the bottom of the sea not being in every place rocky, but here and there sandy, gravelly, and oozy, imbibes the sea-water and letteth it into the earth (after the same manner as when we throw water upon sand, beans, peas, wheat, or other grains) thr' whose interstices it is brought by degrees to a great distance from the sea, where at length the small drops come together, especially in streight places, as are mountains, etc. and having found an aqueduct they discharge themselves at a spring.'[31]

The second question was rather difficult to answer as salt could not be removed by percolation or filtration. The rainfall on the ocean was always of freshwater (elsewhere he said that it was sometimes

Figure 9. A hydrophylacium within a mountain, filled with water, feeds the nearby spring (after Herbinius).

saltish), and hence nature itself had a way of getting rid of salt from the sea-water. If nature can do it in case of precipitation, why can it not be done during the process of percolation? It is also possible that the salt water comes into contact with a vast amount of fresh water, and the subsequent dilution makes it difficult to detect salinity.

Varenius, however, rejected the sea-water hypothesis as the cause of the annual inundation of the Nile, and opted for the precipitation concept.[32]

JOHANN HERBINIUS

The book *Dissertationes de admirandis mundi cataractis supra et subter-*

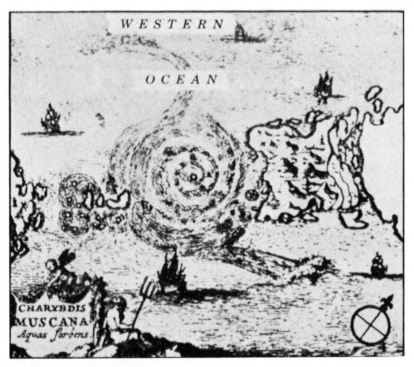

Figure 10. Through a hole in the sea bed in the island of Mosken, Norway, water is sucked in and forced out from the great subterranean reservoir. Water forced out through the hole is responsible for generation of tides (from Herbinius).

raneis by Johann Herbinius was published in Amsterdam, in 1678. Herbinius listed the various probable causes for the origin of rivers and springs as follows: God; continuous movement of water in the subterranean abyss which, by virtue of motion drives the water up to the surface of the earth; angels; stars; the spirit of the earth; and air enclosed within the earth. Strangely enough, Herbinius completely disregarded precipitation as even a possible reason! However, he classified his reasons into two types, true and false; the first two causes were true and all of the others were false. God is the primary cause; the subterranean abyss is the secondary. Figure 9 shows Herbinius' concept of a river originating from a water-filled subterranean cavern.

The Maelström near the island of Mosken, in Norway, particularly attracted his attention.[33] It was:

'no less than 40 miles round, and upon the tides coming in swallowes in a manner the whole sea with an incredible noise, drawing in ships, whales, or whatever else comes within its compass, and dashing them to pieces against the sharp rocks, that there are in the descent of this dreadful Hiatus; and then upon the ebb throwing them out again with a prodigious a violence, in so much that some have attributed the whole flux and reflux of the sea (and not without some reason) to this vast varago.'[34]

The phenomenon (figure 10) served to illustrate his point. The whirlpool indicates an opening in the sea bed, and through it water is sucked in and forced out alternately. The water forced out, he claimed, is responsible for the creation of tides.

BRITISH CONTRIBUTIONS

The major British contributions to hydrology during the seventeenth and the early eighteenth centuries were the evaporation experiments of Edmond Halley, and the rain gauges of Sir Christopher Wren and Robert Hooke. They are discussed in chapters 10 and 11 respectively. With the exception of the above-mentioned scientists, the British contributions to hydrology, in general, tended to be relatively poor in comparison with their achievements in other sciences.

The book *Nature's secrets* by Thomas Willisford, published in 1658, dealt with the subject of hydro-meteorology much in the traditional manner.[35]

John Ray (1627–1705), the English naturalist, achieved most of his fame from his plant classification (which obviously helped the development of systematic botany) but the concept of the hydrologic cycle which he presented also deserves consideration. He believed that sun attracted vapours from the earth and the sea, and that the wind was responsible for driving the vapours from the sea toward the land where it fell as rain. The rivers obtain their supply of water from rain and their interconnection is obvious during the time of flood. A vast quantity of water is carried down to the sea by the rivers annually, and thus, the circulation of water is complete. He did not favour the concept of the subterranean abyss, and commented, 'I hope those who bring up springs and rivers from the great *abyss* will not bring those vapours, which unite into drops and descend in rain from thence too'.[36]

Ray thought that every natural phenomenon was designed by God for a specific purpose. For example, there is more rainfall on earth than necessary to water it because it brings down a great quantity of earth from the mountains or higher grounds which are then spread on lands during floods – thereby rendering them more fruitful.

Thomas Burnet (1635–1715), an Anglican theologian, had a very vivid and brilliant imagination. His pretentious book *Telluris theoria sacra* (The sacred theory of the earth – containing an account of the original structure of the earth and of all the general changes which it hath already undergone or is to undergo till the consummation of all things), published in 1681, contained a fanciful theory of earth's structure. The book was published under the patronage of King Charles II, and was read in all parts of Europe. In chapter 5, he discussed the water of the primitive earth and the origin of rivers and streams. When the earth was first created, there was no ice, snow, hail, or thunder; and,

'as for the winds, they could not be either impetuous or irregular in that earth, seeing there were neither mountains nor any other inequalities to obstruct the course of the vapours; not any unequal seasons, or unequal action of the sun, nor any contrary and struggling motions of the air: nature was then a stranger to all those disorders.'[37]

In the primitive earth, the sun's heat was directed primarily towards the equatorial parts and, consequently, caused most of the evaporation. The vapours raised in this manner were the 'most rarified and agitated; and being once in the open air, their course would be that way, where they found least resistance to their motion; and that would certainly be towards the poles, and the colder regions of the earth.' On reaching colder polar regions, vapours condensed and fell as rain and dew (figure 11). The rain at the poles was continuous due to the vast quantity of water that vapourized in the tropics. Burnet was next confronted with a serious problem: how did water circulate back to the equatorial region as there were no rivers or mountains in his concept of the primitive earth. His answer was rather ingenious. He suggested that the poles had

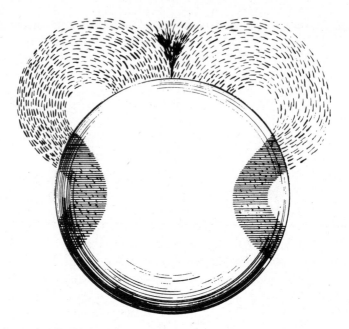

Figure 11. Pattern of evaporation in the primitive earth according to Burnet.

higher elevations than the tropics, and hence, water had to flow downhill and thus form water courses. Hence, the primitive hydrologic cycle consisted of two parts: 'aerial rivers' and surface rivers.

Burnet's work was sharply criticized by Erasmus Warren,[38] the Rector of Worlington, and John Keill.[39, 40] Keill (1671–1721), an 'enthusiastic student of the Newtonian principles', was an able mathematician and astronomer. His objections were precise, and did much to demolish Burnet's fantastic system of air and vapour circulation over the primitive earth. Keill pointed out that an earth without the sea could not possibly supply water vapour to the air. He said:

'And seeing the Sea as it is now laid open to the action of the Sun, is but just sufficient to supply us with Rain and Vapours; does it not seem a thing against common sense to suppose that the Abyss inclos'd with a thick shell could have sent out a quantity of Vapours great enough for such an effect?'[41]

Keill asserted that the sealess earth, instead of being a Garden of Eden as portrayed by Burnet, would be 'nothing else but a Desert'. Keill also disagreed with John Woodward[42] and William Whiston's[43] concept that rivers obtain their source of supply from the subterranean abyss.[44, 45] To him, it seemed almost self-evident that rivers originate from rainfall.

'And as for Rivers, I believe it is evident, that they are furnished by a superior circulation of Vapours drawn from the Sea by the heat of the Sun, which by Calculation are abundantly sufficient for such a supply. For it is certain that nature never provides two distinct ways to produce the same effect, when one will serve. But the increase and decrease of Rivers, according to wet and dry Seasons of the year, do sufficiently show their Origination from a Superior circulation of Rains and Vapours. For if they were furnished by Vapours exhaled from the Abyss through subterraneous Pipes and Channels, I see no reason why this subterraneous fire, which always acts equally, should not always equally produce the same effect in dry weather that it does in wet. . .
I know the maintainers of this Opinion use to alledge, that there are Springs and Fountains on the tops of Mountains, which cannot easily be maintained by a Superior circulation of Vapours: but I beg those Gentlemens pardon, for I can give no credit to any such Observation; for I am well assured, that there are none of those Springs in some places where it is said they are. And particularly that Learned and diligent Observer of Nature Mr. Edward Lloyd the Keeper of the Musaeum Ashmoleanum assured me, that throughout all his Travels over Wales, he could observe no such thing as a running Spring on the top of a Mountain. On these considerations, I think it is not in the least probable, that Rivers and Springs proceed from Vapour, that is, raised by a subterraneous heat through the Fissures of the Mountains.'[46]

Robert Plot (1640–1696), a Fellow of the Royal Society of London, was interested in the origin of streams.[47] Figure 12 shows his

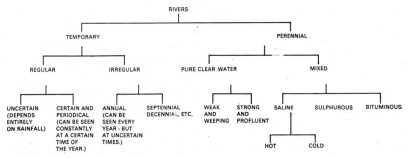

Figure 12. Classification of rivers by Robert Plot.

classification of rivers. He was familiar with Perrault's work but he rejected it, and advocated the sea-water theory. He believed that the salt water ascends to mountain tops through tapering passages that finally become capillaries. The movement of water to an elevation higher than that at which it originally stood, was explained by the pressure of air on the surface of the sea 'just as the quicksilver ascends in the tube of a barometer by the pressure of the air upon the stagnant poole of mercury below'.[48]

In 1695, Dr. John Woodward, Professor of Physic at Gresham College, published a comprehensive book on the natural history of the earth which attracted considerable attention. According to him:

'there is a mighty collection of water inclosed in the bowels of the earth, constituting a huge orb in the interior or central parts of it; upon the surface of which orb of water the terrestrial strata are expanded. That this is the same which Moses calls the Great Deep, or Abyss; the ancient gentile writer, Erabus, and Tartarus.'[42]

The abyss communicates with the ocean by means of hiatuses or chasms. All waters of the earth, springs, rivers, vapours, and rains, are supplied from the standing fund of the reservoir of the abyss with its partner the ocean.[49] He did not agree with Halley's conclusions (see Chapter 10). He suggested that 'springs and rivers do not proceed from vapours raised out of the sea by the sun, borne there by winds unto mountains, and there condensed, as a modern ingenious writer is of opinion'.

Woodward's book did attract several criticisms which were either answered by some of his devoted followers or by the author himself.[50]

ARTESIAN WELLS OF MODENA

Probably the first correct explanation of the nature of artesian wells[51] was given by the philosopher, astronomer, and geographer al Bīrūni (973–1048) of the Arabian school. The same concept was later mentioned casually by Palissy, and both Giovanni Cassini (1625–1712) and Bernardino Ramazzini (1633–1714) paid considerable attention to the subject. Ramazzini, a professor of the medical school at Modena, wrote a book on the wonderful artesian springs of Modena, in 1691.[52] The book so incensed Robert St.

Clair of London that he not only translated the entire work into English but also wrote a lengthy discourse against it.[53] Among his views thereon was the following: 'they come to overturn the Scripture, to establish their own prophane fancies, as our theorist has done, in favour of a spurious brat, of which he will needs be counted the father'.

In his attempt to explain the phenomenon of the artesian wells, Ramazzini said disarmingly that the 'nature of fluid bodies is so abstruse and intricate, that it could never be enough explained by the most solid wits'. The ancient city of Modena has:

'. . . a great abundance of most pure water, which neither can cease through length of time, nor be ever vitiated or diverted by the craft of the enemies: For this city has under its very foundations a great repository of waters, or whatever else it may be called, out of which it draws an inexhaustible stock of waters; and, which is very rare, is got at a very small charge; seeing for the getting of this treasure (for water, according to the testimony of Pindarus, is the best of all things) there is no need of great stir, in digging through mountains, or keeping a great many workmen, as is usual elsewhere, and such as Rome, formerly had divided, as Frontinus says, into searchers, water-finders, water-bayliffs, conveyors, distributors and many other workmen.'[54]

Water was so plentiful in Modena that any citizen could take as much as he liked from the artesian wells without obtaining permission of the local authorities. The general procedure was to dig to a depth of 63 ft, and then to use an auger to bore a hole for the next 5 ft. A vast quantity of water would thereupon gush out with great force. Subterranean conduits were used to convey water to the public canals which fed a navigable canal. Thus, one could travel by boat all the way to Venice from Modena by way of the canal, and the Scultenna and Po rivers.

Ramazzini's explanation of the artesian phenomenon (figure 13) follows:

'First, we may freely affirm, that these waters are not standing, as they are when shut up in a hogshead, but are in continual motion, and that pretty quick: For the noise of that Water which is heard before the perforation in the bottom of the wells does make it manifest enough. . . And I think 'tis probably the matter is so in our fountains, to wit, the water flows out of some Cistern plac'd in the neighbouring mountains, by subterraneous passages, where the earth is firm and hard; but when it has come into the plain, it expatiates far over the sand, and in the way is lifted up to this height when a hole is made with an auger, according to the Laws of Hydrosticks. . .'[55]

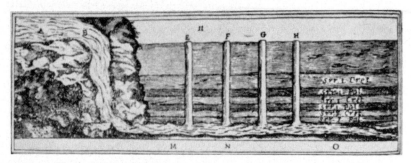

Figure 13. Ramazzini's explanation of the artesian wells of Modena.

'To give some specimen how the flowing of the water may be according to my explication: Suppose, as in figure 10-19 there is a cistern in the bowels of the Appennine, drawing water from the Sea, and that the water is carry'd by sub-terraneous pipes from the same cistern, the water is forc'd to run down by a more narrow space than it had in the beginning, and to follow its course till it come into the Sea, or some great gulph. Therefore wells E F G H being digg'd without any choice in all the tract lying upon this spring, and the hole being made by the auger, the water of necessity must be lifted up on high, being for'd by another, which descending from a higher ground, presses on that which goes before, and drives it up. By this means these waters receive a plentiful supply from their Father Appennine, as does the well of waters which flows from Lebanon, of which there is mention in the sacred history.

But 'tis, by far, more probable, that the Water is sent from the sea into such Cistern, than from Showers, or melted Snows, seeing Rain and Snow-waters run away for the most part by Rivers above Ground; neither can they enter into the ground so deep; as Seneca also testifies.'[56]

'As I have deduc'd from the Original of this Water from the Sea, so I do not deny, that many Fountains owe their Originals to Rains and melted Snow; yet with this difference, that the Fountains, which have their Spring from the Sea by hidden Passages continue perpetual, but those which rise from Showers and temporary Springs at some time of the year, are diminished and quite dry up... Though I derive the Original of our Fountains from the Sea first, then from some Cistern of Water plac'd in our Mountains, into which the Vapors, sent up by the enclos'd Heat, are returned in Form of Waters. I would not thence infer, that this Cistern is plac'd in the tops of the Appennine Mountains, but I believe rather that 'tis plac'd in the Foot of the Mountain, than in the top... But I cannot certainly conjecture in what part, whether near the foot of the Mountain, or in their inner parts this Cistern of Waters is plac'd by the Divine Architect. I have spar'd no Labour nor Experiences to find out the Head of this Spring, and therefore I diligently viewed not only the Plain towards the Mountains, but the Mountains themselves, and could find no Marks of it.

I observed indeed some small Lakes, but such as dry up in Summer, and so become Pasture for Cattle; of the number of which is Lake Paulinus, 25 miles distant from this. I thought best therefore to fetch the Original of these Waters from another source, viz. from some secret Cistern of Water plac'd in the inner parts of the Appennine Mountains. And it is certain, that the inner parts of the Mountains are cavernous, and that there are in them Cisterns of Water, from whence Fountains and Rivers drawn their Original.'[57]

DOMENICO GUGLIELMINI

The Italian, Giovanni Domenico Guglielmini (figure 14), who contended that Mariotte's contribution to the hydraulic sciences was negligible, tried to solve various problems associated with river flows by field observations. His findings had considerable influence on the developments of concepts of open channel flows, not only in Italy but also throughout the world. Born in Bologna, in 1656, he was a professor of mathematics at the university of his native city. His book *Aquarum fluentium mensura nova methodo inquisita*[58] was published in 1690. In 1692, he became the professor of hydrometry at the university in addition to being the water superintendent of the district. His major work on the subject[59] was published in 1697, and the following year he was appointed to the chair of mathematics at Padua. Guglielmini, educated in both medicine and mathematics, never really gave up his practice of medicine as a side line. During the last few years of his life he gave up the study of water sciences, and accepted the chair of theoretical medicine. He died in 1710.

Guglielmini, whom Frisi[21, 22, 60] called a 'great master', was a practical man, and his treatise is primarily on the flow of water in rivers and canals. Toward the late seventeenth century, thanks to Castelli, the continuity equation had become firmly established, but the general principles of open channel flow still needed to be grasped. It was Guglielmini who supplied some of the basic concepts.[61] He conducted experiments on the efflux principle with greater precision than either Torricelli or Maggiotti, and was convinced that the velocity with which water flowed out of an orifice was proportional to the square root of the head. He extended the same analogy to the cases of flow at a sluice gate and at the inlet of a sloping canal. He believed that:

Figure 14. Domenico Guglielmini.

'... water having a certain depth, the upper parts press the lower with the force of gravity and make them move towards any difference of level, which means in practice that each has that exact grade of velocity which it would have acquired descending from the surface to the place where it is located; we must confess that the velocity of water does not depend only on the descent covered on the inclined canal, but also on the pressure exerted by the upper parts upon the lower according to the rule above mentioned.'[62]

Figure 15. Guglielmini's sketch of free-surface configuration.

However, he realized that such a condition could not exist in nature, and reasoned that the parabolic distribution of velocity, with zero velocity at the water surface, would be valid only in case of a perfect fluid.[63]

Leliavsky's claim[64, 65] that Guglielmini believed in the parabolic distribution law has to be discounted. It was probably based on Frisi's statement[60] rather than from Guglielmini's work, and the suspicion is further strengthened because of Leliavsky's immediate discussion of the works of Father Grandi (described in considerable detail by Frisi) and Frisi.

Elementary concepts of open channel flow can be found from the following excerpt:

'Water passing from rest to motion or leaving the reservoir. . . acquires in the descent through the rivers, which are on an inclined plane to the horizontal, some degree of velocity, but this very soon reduces to uniformity (equabilità) due to the great resistances that the water encounters in its motion. . . Once reduced to uniformity, water must, however, maintain the velocity that it acquired previously in flowing on its plane and this is regularly greater, the greater the slope of its bottom.'[66]

The velocity of flow will increase with further increase in flow as the relative effect of resistances will be less, provided it remains constant. Thus, even though the concept was fundamentally correct, it was not expressed quantitatively. Guglielmini's sketch of a free-surface configuration is shown in figure 15.

Guglielmini's reasoning can possibly be extended to obtain the functional relationship between discharge and depth, though the concept was not exactly correct. He said that velocity was proportional to the slope as well as the square root of the depth, which meant that discharge was proportional to depth(3 : 2), but because of his faulty reasoning he would not receive the credit for the concept. The Italian, however, should be credited for his reasoning that the velocity of flow is dependent on the cross-sectional area of the channel and inversely proportional to the wetted perimeter.

NITRE THEORY OF NILE

Toward the latter half of the seventeenth century, a new theory was advanced regarding the annual inundation of the river Nile. It seems to have originated from one Vanslebius who lived in Egypt for some years and who had carefully observed the same phenomenon. The theory was believed by Robert Plot[67] and N. C. De la Chambre,[68] and both of them elaborated on Vanslebius' original concept. According to the hypothesis, certain kinds of drops, somewhat similar to dews, fell in the valley a few days before the annual rise of the Nile, and fermented the river water. Rainfall was categorically rejected as a reason because, according to the theory's originators, the rains do not usually start till the 25th of June whereas the inundations commence as early as the 12th of that month.

The fermentation causing the rise was said to produce a kind of green scum on the water surface. Plot agreed with Cambracus and Gassendi that the fermentation was caused by the presence of nitre.[67] Plot asserted, on the authority of Vanslebius, that the inundation of the Nile was not an isolated phenomenon as water levels in wells in certain parts of Egypt have been observed to

'rise the very same night, and in the same manner with the river, which having no possible communication with the rains in Habessia, shows evidently that the increase of the water in the river, comes partly at least from another cause, and

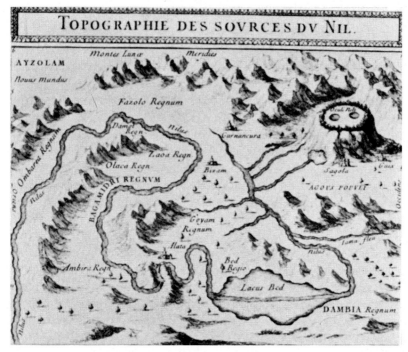

Figure 16. A map of the source of the Nile indicating the presence of nitre pits (after N.C. De la Chambre).

most likely from the fermentation made by the niter, which however it comes to pass seems to leave its owne parts (whereof there are abundance in Egypt) at the time of the increase and goes into the river.'[67]

N. C. De la Chambre[68] believed that the drops also made the mud of the Nile heavier than before, and the total rise of the river in any year could be predicted by the relative weight of the mud. The special mud did not create any problem of sedimentation because the nitre was volatile and, consequently, the heights of the banks had remained constant for more than 2000 years (from the time of the Greek historian, Herodotus). Figure 16 is a topographical map of the source of the Nile according to De la Chambre, on which the locations of nitre pits are shown.

ORIGIN OF STREAMS

The cause or causes of the origin of springs was an extremely popular subject in the seventeenth century, and practically every man of importance expressed his opinion on it. Some of the opinions have already been discussed, and other significant ones will now be mentioned briefly.

Molina (1536–1609) correctly suggested that rivers originated from rainfall which penetrated the earth's crust, but the idea was opposed by Riccioli who held that precipitation could not percolate to more than 12 to 15 feet and much less so in the rocky countries of Peru and Chile where streams are abundant.[69] He quoted from *Genesis:* 'The Lord had not yet rained upon earth', but 'A fountain ascended from the earth which watered its entire surface'.* The subterranean condensation theory was disregarded as it was considered to be too slow a process. According to his calculations, the Volga alone carried enough water to the Caspian Sea every year to inundate the entire earth – a conclusion that was later quoted by Plot in support of his own theory. Riccioli stated that all good Catholics should accept the explanation in Ecclesiastes.[70]

Dobrzensky[71] opposed the pluvial theory on the same grounds as Riccioli, and he suggested two other causes. The remote cause was the tides of the sea, and the local one was the subterranean condensation of air.

Isaac Voss categorically denied the sea-water conversion theory. He believed that all rivers originated from rainfall, but his credibility is somewhat marred by the ridiculous ideas he expressed on other subjects. Among others who expressed correct opinions were Fabri,[72] Bartholinus[73], and Robert Hooke.[74]

In the treatise on 'Running waters', Carlo Fontana,[75] an Italian, stated that the sea must be higher than the highest mountains as no landmarks are visible if one, at sea, is a few miles off shore. He obviously failed to consider the curvature of the earth! The hypothesis, however, provided an excellent base for explaining the movement of water from the sea to the mountain tops: after all, water was not rising above its own level!

---

* Genesis II, 5–6, 'Nondum pluerat Dominus super terram. . . Fons ascendebat a terra irrigans universam superficiem terrae'.

ADVENT OF CURRENT METERS

Most works in the field of hydrometry credit Henry Pitot and Reinhard Woltman with the earliest development of sophisticated velocity-measurement devices during the eighteenth century. The earliest current meter, however, was probably constructed[76] by an Italian physician, Dr. Santorio Santorio (1561–1630), around 1610. Santorio became interested in the subject in a curious way; he wanted to evaluate the critical velocities under which his patients were lulled to sleep by the soft noise of quietly falling water as contrasted with the loud noises produced by rapidly falling waters which would keep them awake. Hence, he proposed:

'. . . as a means of subtly ascertaining the reason for those circumstances, that one should weigh, with a balance scale, the amount of impact produced under both [circumstances]. When urged by friends to show how that could be done, I prepared two staters [Roman balances] on both of which a plate is fixed at right angles. By means of the first [with the plate pointing upwards], the impact of winds could be weighed; by means of the second [with the plate pointing downwards], the impact of the water. The one having the plate pointing upwards for measuring the impetus of the winds helped us to predict when such winds tended to increase. We could thus foresee heavy weather at sea, and could avoid the perilous sinking of ships. It at least provided a means whereby we could be more sure whether the impact of winds tended to increase or decrease. Other uses for this device are described in the book regarding medical instruments. The other instrument, on which the same [size] plate hangs downward, allows us to weigh the amount of impact produced by running waters. It might have primary usefulness for improving the efficiency of [water] mills, and for many other applications. In consequence of its use, we shall be able to ascertain what amount of impact has beneficial [medical] properties, and what amounts would have harmful properties. Actually, if certain measures of impetus or noise are salubrious, and others are unsalubrious, by what better method could we graduate the strengths of the medicines we are to take in.'[77]

Santorio's drawings of the two instruments are shown in figures 17 and 18.

The instrument only measured the impact of water striking the plate, and no effort was made to convert the data into water velocities. In fact, the mathematical tools for making such conversions had not yet been conceived.

Robert Hooke, the inventive genius, also designed an early current meter. An entry in the volume VI of the *Register book* of the Royal

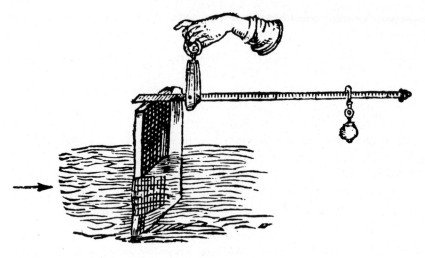

Figure 17. Santorio's water current meter.

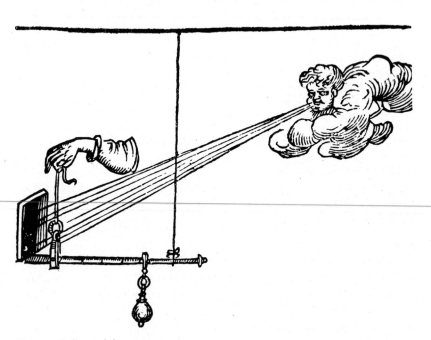

Figure 18. Santorio's anemometer.

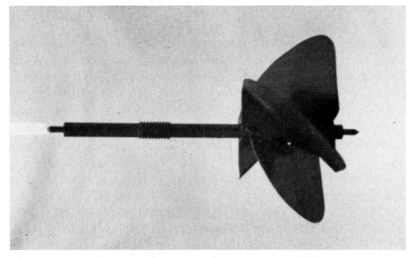

Figure 19. Model of the rotor for a sounding apparatus of Robert Hooke (reconstructed by Arthur H. Frazier).

Society of London entitled 'A way-wiser* for sea' and dated November 28, 1683, reads:

'I shewed an Instrument I had contrived and shewed some of the Society about 20 years since, By which the way of a Ship through the Sea might be exactly measured as also the Velocity of any running Water or River and thereby the comparative velocity of it in its several parts, by this also the quantity of the water vented by any River into the Sea or any other River might be found.'[78]

Unfortunately, no sketch of Hooke's current meter seems to have been preserved. Frazier, however, has reconstructed[79] its rotor from other descriptions (figure 19). He comments that:

'the facilities Woltman provided for counting the revolutions are identical with those which Hooke had provided; and if Hooke's rotor had been made in an appropriate size, and were mounted on a frame such as that which Woltman had adopted, it could probably have served the purpose even better than Woltman's.'[79]

---

* A ship's log in this case.

CONCLUSION

Looking back, the greatest achievements of the seventeenth century, omitting the contributions of Perrault, Mariotte and Halley, were the restatement as well as the general acceptance of the continuity principle. The credit for it must go to Benedetto Castelli, even if he obtained the idea from Leonardo (which is debatable). One has to applaud him for his contribution to make the principle widely known and, what is more important, almost universally accepted. The resurgence of interest in practical hydrologic engineering was obviously due to the importance of irrigation, river-training, and flood control projects. The construction of such works gave rise to innumerable problems that could only have been successfully solved by a better insight to the physical phenomena involved. Probably if the responsible engineers of that day encountered a particular problem, they would first turn to the Greek and the Roman masters for an answer, but in most instances they would have been disappointed. Thus, they had to find their own solutions and this circumstance alone was a significant step forward. The hydrologic field was dominated by Italians in the seventeenth century, notably by Castelli, Torricelli, Ramazzini, and Guglielmini. This circumstance extended well into the eighteenth century with the works of engineers like Vallisnieri, Manfredi, Grandi, Frisi, and Poleni. The explanations of Ramazzini regarding the artesian wells of Modena and the observations of Guglielmini on the open channel flow also deserve special mention.

REFERENCES

1. BACON, R., Novum organum, Book 1, XIX, quoted by A. R. Hall, The scientific revolution (1500–1800). London, Longmans, Green & Co. (1954) p. 161.
2. LENOBLE, R., The 17th century scientific revolution. In: The beginnings of modern science, edited by R. Taton. London, Thames & Hudson (1964) p. 190.
3. CUVIER, G., Essay on the Theory of the earth, translated by Prof. Jameson, 3rd ed. Edinburgh, William Blackwood (1817) pp. 44–45.
4. BISWAS, ASIT K., The hydrologic cycle. Civil Engineering, ASCE 35 (1965) 70–74.

5. MEINZER, O. E., The history and development of ground-water hydrology. Journal of the Washington Academy of Sciences *24* (1934) 10–11.
6. SCOTT, J. F., The scientific work of René Descartes. London, Taylor and Francis Ltd. (1952) p. 67.
7. KEILHACK, K., Lehrbuch der Grundwasser- und Quellenkunde. Berlin, Gebr. Borntraeger (1912) pp. 76–77.
8. MARTEL, E. A., Nouveau traité des eaux souterraines. Paris, G. Doin (1921) p. 78.
9. PATRIN, M., Springs. In: New dictionary of natural history, published in many volumes (1816–1830).
10. CASTELLI, B., Della misura dell'acque correnti. Roma, Nella Stamperia Camerale (1628).
11. BISWAS, ASIT K., Discussion of: Ground-water management for the nation's future – ground-water basin management. Journal of the Hydraulics Division, ASCE *91* (1965) 382–384.
12. CASTELLI, B., Of the mensuration of running waters, translated by T. Salusbury. London, W. Leybourn (1661) p. 5.
13. *Ibid.*, p. 38
14. *Ibid.*, pp. 9–12
15. FONTANA, G., Dell'accrescimento che hanno fatto i fiumi, torrenti e fossi che hanno causato l'inondatione a Roma il Natale. Gioisoi, Appresso Antonio Maria (1640).
16. WOLF, A., A history of science, technology, and philosophy in the 16th and 17th centuries. London, George Allen and Unwin Ltd. (1935) p. 540.
17. LOMBARDINI, E., Dell'origine e del progresso delle scienza idraulica nel Milanese ed in altre parti d'Italia. Milano, Tip. di D. Salvi e Comp. (1860) p. 35.
18. *Ibid.*, pp. 40–41.
19. POGGENDORFF, J. C., Geschichte der Physik. Leipzig, J. A. Barth (1879).
20. MIDDLETON, W. E. K., The history of the barometer. Baltimore, Johns Hopkins Press (1964).
21. FRISI, P., Del modo di regolare i fiumi e torrenti. Lucca (1762).
22. FRISI, P., A treatise on rivers and torrents, translation of ref. 21 by J. Garstin. London, Longmans, Hurst, Rees, Orme and Brown (1818) pp. 43–50.
23. KIRCHER, A., Mundus subterraneus. Amstelodami, Apud J. Janssonium et E. Weyerstraten (1664–1665).
24. DE WIEST, R. J. M., Geohydrology. New York, John Wiley and Sons Inc. (1965) p. 6.
25. KIRCHER, A., The vulcanos: or, burning and fire-vomiting mountains, anonymous translation of part of: Mundus subterraneus. London, J. Darby for John Allen (1669).
26. KIRCHER, A., Itinerarium exstaticum, quo mundi opificium, typis V. Mascardi Romae (1656).
27. SCHOTT, R. G., Anatomia physico-hydrostatica fontium ac fluminium. Herbipoli (1663).

28. ADAMS, F. D., The birth and development of the geological sciences. New York, Dover Publications Inc. (1954) pp. 443–444.

29. BECHER, J. J., Chymisches Laboratorium, oder untererdische Naturkündigung. Franckfurt, J. Haasz (1680).

30. VARENIUS, B., Geographia generalis. Amstelodami, L. Elzevir (1650).

31. VARENIUS, B., A compleat sytem of general geography, translated by Mr. Dugdale, 2nd ed. London (1734) pp. 298–314.

32. Ibid., pp. 330–338.

33. HERBINIUS, H., Dissertationes de admirandis mundi cataracti ssupra et subterraneis. Amstelodami, Janssonio-Waesbergios (1678) p. 129.

34. PLOT, R., The natural history of Staffordshire. Oxford, The Theatre (1686) p. 73.

35. WILLISFORD, T., Nature's secrets or the admirable and wonderful history of the generation of meteors. London (1658) pp. 38–39.

36. RAY, J., The wisdom of God manifested in the works of creation. London (1691).

37. BURNET, T., The theory of the earth, 2nd ed. London (1691) pp. 221–228.

38. WARREN, E., Geologica. London (1690).

39. KEILL, J., An examination of the reflections on the Theory of the earth. Oxford, The Theatre (1698).

40. KEILL, J., An examination of Dr. Burnet's Theory of the earth, together with some remarks on Mr. Whiston's New Theory of the earth, 2nd ed. London (1734).

41. Ibid., p. 68.

42. WOODWARD, J., An essay toward natural history of earth and terrestrial bodies. London (1695) pp. 117–118.

43. WHISTON, W., A new theory of the earth. London (1696).

44. WHITE, G. W., John Keill's view of the hydrologic cycle, 1698. Water Resources Research 4 (1968) 1371–1374.

45. TUAN, YI-FU, The hydrologic cycle and the wisdom of God. Toronto, University of Toronto Press (1968).

46. KEILL, J., Ref. 40, pp. 187–188.

47. PLOT, R., De origine fontium tentamen philosophicum. Oxonii (1685).

48. PLOT, R., Ref. 34, p. 79.

49. WOODWARD, J., Review of ref. 42 in Philosophical Transactions of the Royal Society of London 19 (1695) 115–123.

50. WOODWARD, J., A supplement and continuation of the Essay toward natural history of the earth, translated by B. Holloway. London (1726) pp. 96–100.

51. DAMPIER, W. C., A history of science. Cambridge, University Press (1948) p. 75.

52. RAMAZZINI, B., De fontium mutinensium admiranda scaturigine tractatus physico-hydrostaticus. Mutinae (1691).

53. ST. CLAIR, R., The abyssinian philosophy confuted or Telluris theoria, neither sacred nor agreeable to reason, translation of ref. 52, and a lengthy discourse against it. London (1687).

54. *Ibid.*, p. 1.
55. RAMAZZINI, B., *op. cit.*, p. 29.
56. *Ibid.*, p. 130.
57. *Ibid.*, pp. 130–144.
58. GUGLIELMINI, G. D., Aquarum fluentium mensura nova methodo inquisita. Bononiae (1690).
59. GUGLIELMINI, D., Della natura de' fiumi, trattato fisico-mathematico. Bologna (1697).
60. MASOTTI, A., Scritti inediti di Paolo Frisi. Istituto Lombardo di Scienze e Lettere *78* (1944–1945).
61. GUGLIELMINI, D., Opera omnia mathematica, hydraulica, medica et physica, vol. 1, edited by J. B. Morgagni. Ginevra (1719) pp. 105–726.
62. LOMBARDINI, E., Dell'origine e del progresso della scienza idraulica. Milano (1872).
63. ROUSE, H. and S. INCE, History of hydraulics. Iowa City, Iowa Institute of Hydraulic Research (1957) p. 69.
64. LELIAVSKY, S., Historic development of the theory of the flow of water in canals and rivers, no. 1. The Engineer *91* (1951) 466–468.
65. LELIAVSKY, S., River and canal hydraulics. London, Chapman and Hall (1965) p. 4.
66. DE MARCHI, G., Guglielmini. Brescia (1917).
67. PLOT, R., Ref. 34, pp. 54–56.
68. DE LA CHAMBRE, N. C., Discours sur les causes du débordement du Nil. Paris (1665).
69. RICCIOLI, G. B., Geographiae et hydrographiae reformata libri duodecim. Bononiae, V. Benatiij (1661).
70. RICCIOLI, G. B., Almagestum novum. Bononiae, V. Benatiij (1651) p. 68.
71. DOBRZENSKY, J. J. W., Nova et amoenior de admirando fontium genio. Ferrariae, Apud A. et J.-B. de Marestis (1657) pp. 1–9.
72. FABRI, HON., Physica, vol. 3. Lugduni (1670) p. 393.
73. BARTHOLINUS, C., Speciminis philosophiae naturalis novissimis rationibus et experimentis, ch. 13. Hafniae (1690).
74. BIRCH, T., The history of the Royal Society of London, vol. 3. London (1757) p. 221.
75. FONTANA, C., Utilissimo trattato dell'acque correnti. Roma, G. F. Buagni (1696) p. 9.
76. FRAZIER, A. H., Dr. Santorio's water current meter, circa 1610. Journal of the Hydraulics Division, ASCE *95* (1969) 249–254.
77. SANTORIO, S., Commentaria in primam fen primi libri canonis Avicennae. Venetiis, Sarcinam J. Sarcinam (1625).
78. REGISTER BOOK, vol. 6. London, Royal Society of London (manuscript).
79. FRAZIER, A. H., Robert Hooke and the Hooke current meter. Journal of the Hydraulics Division, ASCE *95* (1969) 439–446.

10

Beginning of quantitative hydrology

INTRODUCTION

Towards the end of the seventeenth century, nearly all investigators of distinction belonged to one scientific society or another. The most notable of those societies were the Royal Society of London, and the Académie Royale des Sciences of Paris, founded in 1662 and 1666 respectively. Publishing the results of new investigations in journals of the societies was a logical occurrence, and the two societies gradually attained a prestige greater than that of most universities.

Three men of this period are particularly outstanding for their contributions to the development of the science of quantitative hydrology. All three were connected, directly or indirectly, to both of those societies. They were the French naturalist, Pierre Perrault; the French physicist, Edmé Mariotte; and the English astronomer, Edmond Halley. As quantitative hydrologists, they were the first to undertake experimental investigations to establish some of the fundamental principles in that science. Mariotte and Halley, members of the Académie Royale des Sciences and the Royal Society respectively, played significant parts during the early years thereof, and although Perrault did not belong to either of those scientific societies, his brother, Claude, was an active member of the French Academy, and Christiaan Huygens (1625–1695), the physicist, to whom his book *De l'origine des fontaines* is dedicated, was considered a mainstay of the Academy. In any event these three scientists, Perrault, Mariotte, and Halley, for the first time in history, started the science of hydrology off on a quantitative basis.

PIERRE PERRAULT

Only a little is known about the personal life of Pierre Perrault (1608–1680). Following his father's footsteps, he became a lawyer. He bought the post of receiver-general of finances for Paris, but found it necessary to help himself from the treasury to satisfy his creditors. He was caught while performing that offense, and was dismissed by Colbert who was then at the height of his power. Pierre was rather overshadowed in his time by his three younger brothers: Claude (1613–1688), a physician, scientist and the architect of the Louvre Museum; Nicholas (1624–1662), a noted theologian; and Charles (1628–1703), a critic and the author of Mother Goose fairy tales. Claude became an important figure in the history of science, and was undoubtedly one of the most eminent scholars of his time. In 1673, he translated Vitruvius' work on Architecture.

The book, De l'origine des fontaines[1] (The origin of springs), was published anonymously in Paris, in 1674, and was dedicated to 'Monsieur, Mr. Huguens de Zulichem'. The authorship of the book has been the subject of considerable controversy in the past, and has been attributed variously to André Félibien, Denis Papin, and finally to its true author, Perrault. The first originated with an erroneous classification in the catalogue of books in the Aguesseau Museum (No. 3297); and the second in the Philosophical Transactions of the Royal Society, where a review of the book first appeared under the following heading: 'A particular account given by an anonymous French Author in his book on the Origin of fountains printed in Paris in 1674 to show that the Rain and Snow Waters are sufficient to make Fountains and Rivers run perpetually'.[2] That review appeared in 1809 in an abridged edition of the Transactions where the name of the author was given as Denis Papin. Figure 1 is reproduced from the original 1674 edition. The work is presently attributed universally to Perrault,[3, 4] and is so listed in the catalogues of the British Museum, the Bibliothèque Nationale, and the Library of Congress. In a further confirmation thereof, it was found that N. A. Félibien had written Perrault's

Figure 1. The frontispiece from Perrault's *De l'origine des fontaines*.

name as its author in a first edition thereof which had been presented to him 'Du don de l'autheur, ce 23 octobre 1674'. That copy, incidentally, was sold in Paris in 1872.[5] An excellent investigation into the authorship of the book has been published in an article by Dooge.[6] Perrault's book on the origin of springs was initially written just for private circulation among his friends, but because of its importance he was persuaded to have it published. In it he reviewed the various propositions of the previous authors on the subject (demolishing their conclusions in most cases), and he proposed that an experimental investigation be performed to prove that rainfall alone is sufficient to support the flow of springs and rivers throughout the year. Parts of the book had been translated previously,[2, 6, 7] but it was not until 1967 that the entire work was brilliantly translated with annotated references, by Aurèle La Rocque.[3]

Perrault's proposition was simple:

'It would be necessary, to attain our goal, to measure or estimate the water in some river as it flows from its source to the spot where some brook joins it, and see if the rain water that falls around its bed when put into a reservoir, as Aristotle says, would be sufficient to cause it to flow for a whole year. I have seen the Seine River, and have examined it rather closely in its course from its source to Aynay le Duc, where a brook joins and enlarges it: that is why I shall take it as the subject of the examination I wish to make.

The course of this young River from its source to Aynay le Duc is about three leagues long, and the slopes of its bed extend to right and left about two leagues on each side, where there are other brooks that flow elsewhere; and since these brooks need rain water for their subsistence as much as the Seine, I wish to count only one half of this space of the sides, and say that the area where the Seine flows, is from its source to Aynay le Duc three leagues long by two leagues wide, and then I shall argue as follows. [The catchment area studied is shown in figure 2.]

If a reservoir had been made with this length and width it would be six square leagues in areas, which when converted to toises [about 6 ft] according to the measurements already established, would make 31,245,144 toises in area.

Into this reservoir we must imagine that rain has fallen for one year to the height of 19$^1/_3$ inches* which is the height of an average year, as we have noted. This height of 19$^1/_3$ inches gives 224,899,942** muids of water or approximately, according to the measure we have agreed upon.

---

* The French inch (pouce or poulce) was divided into 12 lines, and was equal to 27 mm.
** The muid of Paris was equal to 24 hectolitres for salt, and to 268 litres for wine. The latter is probably closest to Perrault's measure.

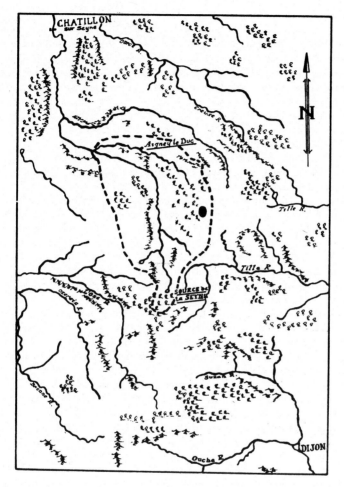

Figure 2. The catchment area studied by Perrault (by courtesy of James C. I. Dooge).

All this water thus accumulated in the quantity just mentioned is what must be used to cause this River to flow for one year, from its source to the place designated, and which must serve also to supply all the losses, such as the feeding of trees, plants, grasses, evaporation, useless flows into the River which swell it for a time and while it rains, turning away of waters which can take another course than that towards this River because of irregular and opposite slopes, and other such wastes, losses, and reductions.

As to the measurement or estimation of the water of this young river, it would be hard to find it exactly and to say what quantity it furnishes: Nevertheless as far as I have been able to judge it cannot have more than 1000 to 1200 inches always flowing, compensating for less at its source with more which it has towards Aynay le Duc, which I judge by the comparison I make of these waters with those of the Gobelins River as it is near Versailles where it has 50 inches of water according to measurements made of it: thus I estimate that it will be enough to give 24 or 25 times as much to ours: for its bed is only four or five toises wide, its depth is small, it will not float boats, and serves only to carry logs which are thrown into it singly to be tied together lower down and to make them into floating log rafts.

Having thus supposed all these things, I say that according to the measurements we have agreed upon, 1200 inches of water give in 24 hours, on the basis of 83 muids of water per inch, 99,600 muids of water; and during one year which is 366 times as much, they will give 36,453,600 muids. This River therefore carries within its banks from its source to Aynay le Duc during one year only the said quantity of 36,453,600 muids of water. But if I draw this quantity of water from the 224,899,942 muids, which are in this reservoir which we have just imagined, there will still be left, 188,446,342 muids, which amounts to almost five times as much, and serves for the losses, decreases and wastes which we have noted. Only about one-sixth part of the rain and snow water that falls is therefore needed to cause this river to flow continually for one year.

I know very well that this deduction has no certainty: but who could give one that would be certain? Nevertheless whatever this one may be, I believe it is more satisfactory than a simple negative like Aristotle's and that of those who maintain, without knowing why, that it does not rain enough to supply the flow of rivers. However that may be, until someone makes more exact observations, by which he proves the contrary of what I have advanced; I shall hold to my view, and be content with the feeble light shed by the observation I have made, being unable to have a stronger one.

If then these waters are enough for the flow of one River, they will suffice for all the other Rivers of the World in proportion, having regard mainly to what is left for wastes, which is more than enough, and to the little space I have allotted on each side of the bed for the River which is only one league on each side: for Rivers are not usually only two leagues apart. It is therefore likely to say, that the waters of rains and snows are sufficient to cause the flow of all the Rivers of the World.'[8]

Perrault, however, did not believe in general infiltration of rain water and thereby recharge of ground water. He went to great lengths to find evidence of general infiltration, and from his observations he concluded that it was an occasional and local phenomenon. In the beginning of the second part of this book, Perrault differentiates his own view from that held by Vitruvius, Gassendi, Pallisy and Francois which he called 'general opinion'. He objected to their concept of infiltration of rain water into the earth, and said that he did not believe that there was just enough precipitation to soak the earth as much as necessary, and then a sufficient quantity was left over to cause rivers and springs. Thus, the two opinions may be compared as follows:

general opinion: rain → infiltration → springs → rivers,
Pierre Perrault: rain → rivers → springs.

EDMÉ MARIOTTE

Although a strong school in water sciences developed in Italy towards the latter half of the seventeenth century, Edmé Mariotte, one of the few outstanding men of the time, did not belong to it. He was probably the most eminent hydrologist of the pre-eighteenth century era. Regrettably, little is known about his personal life, and even the place and date of his birth are uncertain.[9] Since the mathematician Condorcet[10] was unable to find any information about Mariotte's life within the first hundred years after Mariotte's death, our chances of doing so now, some three centuries after his death, seem remote. We must be content to judge him merely from his writings.

Mariotte seems to have been born in Dijon in 1620, and spent most of his life there. In his early youth he was destined for the clergy, and in 1666, when Colbert founded the Académie des Sciences, Mariotte was a prior in the monastery St. Martin-sous-Beaune near Dijon. He became one of the first members of the Academy, and was an active member at that, seldom missing a meeting. He pursued scientific investigations into a great variety of subjects, with great zeal, which suggests a probability that he could not have spent much time at his duties as a prior. He presented papers to the Academy on many subjects including notes of the trumpets, the recoil of guns, the nature of colour, and motion of fluids. Perhaps his greatest contribution to science was the Boyle–Mariotte law, which states that the volume of a given mass of gas varies inversely with the impressed pressure.

His works on hydraulics and hydrostatics, presented before the Academy in 1669 and published posthumously in 1717, had a considerable bearing on the later development of hydraulics and hy-

Figure 3. The Académie Royale des Sciences at work in the Royal Library, Paris, in 1671. Edmé Mariotte is one of the central figures in skullcap and spectacles (from a vignette by Sébastien Le Clerc).

Figure 4. Enlargement of the central group in figure 3. From left to right are Louis Gayant, Edmé Mariotte, Claude Perrault and Jean Pecquet.

drology. During the early days of his final illness, Mariotte gave the manuscript of his book *Traité du mouvement des eaux et des autres corps fluides* to his learned friend Phillipe de la Hire. De la Hire was given full responsibility to finish and edit the manuscript, and he did so admirably. The book was published in 1686. It was translated into English, in 1718, by Rev. J. T. Desaguliers, one of the several men who supported the syphon theory for the ebbing and flowing of springs.

Mariotte died in Paris on May 12, 1684 – the year in which he had clearly demonstrated by his experimental investigations that rainfall was the source of water discharged by springs and rivers.[11] According to Condorcet 'the upright and disinterested love of truth displayed by Mariotte is excelled by very few investigators',[10] and Rühlmann, in his book on mechanics,[12] expressed the belief that Mariotte's work presented the first worthwhile measurements of velocity of flowing water.

Figure 3 is a copy of probably the only surviving picture of Mariotte.[13] The enlargement of the central portion (figure 4) shows four distinguished men of the era – Louis Gayant, Edmé Mariotte, Claude Perrault, and Jean Pecquet.

Treatise on the motion of water

Mariotte's book on the motion of water and other fluids[14–16] was divided into the following five parts:

(1) several properties of fluid bodies, the origin of fountains, and the causes of winds;

(2) the equilibrium of fluids by their gravity, impulse and spring;

(3) running and spouting water and their measurements;

(4) the heights of perpendicular and oblique jets and of their amplitudes; and

(5) the conveyance of water and the resistance of pipes.

Of special interest to hydrologists are the causes of origin of springs (part 1, chapter 2), the determination of velocities of running water at the surface and at the bottom (part 2, chapter 3), and the measurement of discharge in a river or an aqueduct (part 3, chapter 4).

Origin of springs

Mariotte claimed that rainfall was more than adequate for the creation of springs and rivers, and in this concept he was probably influenced by Perrault's work. Percolation takes place after precipitation. When rainfall occurs on hills and mountains, it penetrates the surface of the earth, particularly when the soil is light and mixed with pebbles and roots of trees. Then, if it encounters a layer of clay or a bed of continuous rock, which it is unable to infiltrate, it flows along the surface thereof until it finds a point of egress either at the bottom of the mountain or at a considerable distance below the top. There it breaks out as a spring. Besides proving by experimental investigations that the amount of water that falls as rain is more than enough to supply all the springs and rivers, he demonstrated that the increase or decrease in the flow of springs is directly related to the amount of precipitation. If it does not rain for two months, most streams lose half of their flow, and when a rainless condition continues for a year, most of them become dry. The few which continue to flow do so with a greatly reduced discharge.

Mariotte did not believe in the hypothesis about condensation taking place in vast vaults within mountains, and explained why such an occurrence was impossible:

'For if ABC [figure 5] is a vault in the mountain DEF; it is evident, that if the vapours shou'd become water in the concave of the surface ABC, that water would fall perpendicularly towards HGI, and not towards L, or M, and consequently would never make a spring: Besides, it is deny'd that there are many such hollow places in mountains, and it can't be made appear that there are such. If we say there is earth on the side of, and beneath ABC, it will be answered, that the vapours will gush out at the sides towards A, and C, and that very little will become water; and because it appears that there is almost always clay where there are springs, it is very likely that those supposed distilled waters can't pass through, and consequently that springs can't be produc'd by that means.'[17]

Mariotte measured the total annual rainfall at Dijon as part of his demonstration showing that rainfall was adequate to supply the water flowing in streams and rivers. The observations were taken very skillfully according to his directions, and the annual precipitation was found to be 17 in. After comparing it with the total of 19 in., $2^{1}/_{3}$ lines, observed by Perrault, he selected a conservative value of 15 in. for use in his calculations. Keilhack[18] states that

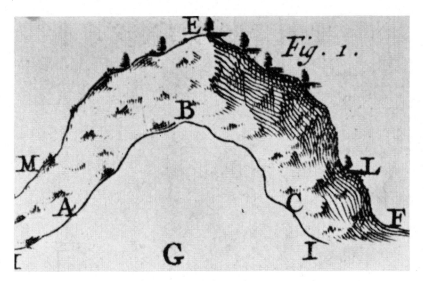

Figure 5. Mariotte's sketch of a mountain vault.

Mariotte had determined the rainfall at Dijon for three years, and had found the average annual rainfall to be 19 in., $2^1/_3$ lines (the same result that Perrault had obtained). A possible explanation for that seemingly strange coincidence, as Dooge[6] has pointed out, is that the value came from a misreading of Mariotte's book, rather than from a knowledge of Perrault's experiments.

Mariotte's calculation was:

15 in. of annual rainfall　　= 45 cu. ft water/square fathom/year
Assume 1 league　　　　　= 2300 fathoms,

Therefore rainfall per square league per year is

$$= 2300 \times 2300 \times 45$$
$$= 238,050,000 \text{ cu. ft.}$$

The catchment area of the river Seine near Paris (figure 6) was assumed to be 60 leagues long by 50 leagues wide. Thus the total annual rainfall on the catchment area was:

$$= 60 \times 50 \times 238,050,00$$
$$= 714,150,000,000 \text{ cu. ft.}$$

The discharge in the river Seine was determined near Pont Royal. The width of the river was taken to be 400 ft, but as the depth varied from 10 ft to 2 ft, the mean depth was taken to be 5 ft. Mariotte

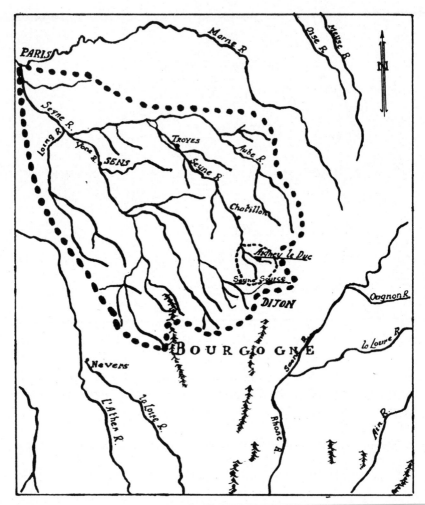

Figure 6. The catchment area studied by Mariotte (by courtesy of James C. I. Dooge).

found that 'when the waters were at their greatest height', a float placed in the middle of the river had the same velocity as a man walking very fast which was equivalent to 15,000 ft per hour or 250 ft per minute. The velocity at the mean depth was found to be 150 ft per minute, and since 'the bottom of the water does not go so swiftly as the middle, nor the middle so fast as the upper

surface', the mean velocity was taken to be 100 ft per minute.
Cross-sectional area of the Seine:

400 × 5   = 2000 sq. ft.

Volume of water passing per minute:

2000 × 100   = 200,000 cu. ft.

Thus the total volume of water passing in the Seine, near Pont
Royal, per year was 105,120,000,000 cu. ft. This was less than
one-sixth the total amount of precipitation in the catchment area.
If an annual precipitation of 18 in. were used instead of 15 in., the
105,120,000,000 cu. ft of annual runoff to 856,980,000,000 cu. ft. of
annual rainfall.

By similar calculations he showed that the total annual discharge
of the great spring at Mont-Martre was about one-fourth of the
yearly precipitation.

Discharge determination

Like Leonardo da Vinci and Benedetto Castelli, Mariotte used
floats to determine velocities in open channels. He suggested that a
ball of wax be used as a float with something heavy inside to sub-
merge it in water as deeply as possible, without sinking. The
primary reason for such a measure was to reduce as much as possible, the affect of wind on its travel. The speed of the float over a
reach of 15 ft to 20 ft was timed with a half-second pendulum. To
calculate the discharge, he suggested that one 'multiply the breadth
of the aqueduct by the height of the water, and the product by the
space which the wax shall have run thro'. The last product which
is solid, will give all the water which shall have pass'd during the
time of observation'.[19] He emphasized that such calculations assume
that the velocity of water to be the same at its top, bottom, and sides.
Probably Mariotte was the first man to realize that water surface
should have the same inclination as the river bed if accurate results
are to be obtained.

He presented an example to indicate the manner of calculating
the mean velocity of an aqueduct having a cross-sectional area of
2 ft by 1 ft. The float was carried 30 ft in 20 seconds, but since

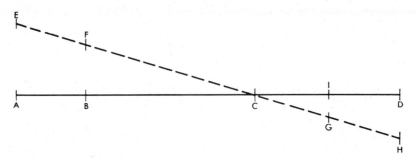

Figure 7.

velocity is greater at surface than at bottom, the mean velocity was
taken to be 20 ft in 20 seconds. It is interesting to note that in all his
calculations Mariotte used a mean velocity which was two-thirds
of the surface velocity. The unit, inch, was also used for expressing
discharge. This was defined as 'the quantity of water which runs
thro' a circular hole of an inch diameter, vertically made in one of
the sides of a vessel, when the surface of the water, which supplies
the running out, always remains at the height of one line above the
hole'.[20]

Discharge of the aqueduct was calculated by:

Cross-sectional area   = $2 \times 1 = 2$ sq. ft
Discharge per 20 seconds   = $2 \times 20 = 40$ cu. ft
                            = $40 \times 35 = 1400$ pints

(1 cu. ft $= 35$ pints of water)

Hence discharge per min.   = $1400 \times 3 = 4200$ pints

$$= \frac{4200}{14} \qquad = 300 \text{ in.}$$

Mariotte used interconnected floats to demonstrate that the velocity
at the bottom was less than at the surface. He used two wax balls
interconnected by a 1-ft string, the lower ball filled with stones, so
that when placed in water it dragged the lighter ball down until the
upper part was just even with the water surface. In a 3-ft deep river,
he found that the lower ball always lagged behind. But in places of
local constrictions, the lower ball travelled faster than the upper
one. He explained the phenomenon as shown in figure 7. Assume
that ABCD (figure 7) was the elevation of the original water surface.

Due to the presence of a constriction near B, water level rises to the dotted line EF. Obviously water would run faster along the steep declivity EFC, and owing to the higher velocity thus attained, it would continue along GH which means that velocity at G and H would be higher than at I and D. This was also the reason, he explained, for the formation of great cavities just downstream from bridge piers.

EDMOND HALLEY

Edmond Halley (figure 8), the eldest child of a rich salter, was born in London in 1656. He was educated at St. Paul's School and Queen's College of Oxford University, but did not finish his course of studies. His first paper (in Latin) on planetary orbits was published in the Philosophical Transactions of the Royal Society before he even reached the age of twenty. He sailed for St. Helena in November 1676, with one of his fellow-undergraduates,[21] to observe the southern stars, having the full consent of his father who gave him a generous allowance of £ 300 per year – more, perhaps, than he ever earned in his life.[22] St. Helena was chosen because it was the southernmost of the English colonies. There he catalogued the latitudes and longitudes of 341 stars, grouped by constellations.[23] He also made numerous pendulum observations and was the first man to record the complete transit of the planet Mercury. On his return from the island, King Charles the Second persuaded Oxford University to grant him the Master's degree without his having residential qualifications or even taking examinations. In 1678, at the age of 22, he was elected a fellow of the Royal Society.

In August 1684, he met Newton, and it was the beginning of a life-long friendship. The following year he resigned the Fellowship of the Royal Society to become its clerk, a post he held till 1698, when he joined the Navy as a Captain to conduct the first British expedition to study Antarctic icebergs and penguins. He was appointed the Savilian Professor of Geometry at Oxford, in 1704, and nine years later, became the Secretary of the Royal Society. At the age of 64, in 1720, he succeeded Flamsteed as the second Astronomer Royal. He was elected a foreign member of the French Academy of Sciences in 1729, and died in 1742.

Figure 8. Edmond Halley, a portrait by Thomas Murray (by courtesy of the Royal Society of London).

Edmond Halley, undoubtedly, was a versatile genius. Primarily known as a pioneer in astronomy, geophysic, and mathematics, he was interested in subjects such as history, archaeology, navigation, and civil engineering. He wrote poems in Latin, translated books from Arabic and Greek, and was the founder of population and actuarial statistics, and a co-founder of experimental hydrology.

Experiments on evaporation
'Men of Gresham' was the name given to the fellows of the Royal Society of that time,[24] whose meeting place was at the Gresham College (figure 9). It was here that Halley conducted his well-known experiments in evaporation. The results were reported in a series of four papers published in the Philosophical Transactions of the Royal Society in 1687, 1691, 1694, and 1715 respectively.

Halley's work on evaporation may well have been motivated by the experiments of Perrault and Mariotte. Both of their books were reviewed in the Philosophical Transactions of the Royal Society[2, 15] and Halley may have decided to try to prove the remaining half of the hydrological cycle, i.e., that enough water evaporates from the oceans and the water courses to produce the rain for replenishing the flow of rivers. Be that as it may, his interest in the evaporation phenomenon was first aroused when he was on the island of St. Helena. His celestial observations were generally taken at night on tops of hills about 2400 ft above sea level. He found, to his great annoyance, that there was such a heavy condensation of vapour when the sky was clear, that he had to wipe his lenses every few minutes and also had trouble in recording his observations because the paper became so wet that ink could not be used. After his return from the expedition, he undertook to explain the 'grand phenomenon' of the equilibrium of the sea 'which is so justly performed that in many hundreds of years we are sufficiently assured that the sea has not sensibly decreas'd by the loss of vapour; nor yet abounded by the immense quantity of fresh water it receives continually from the rivers'.[25]

Halley's explanation of the process of evaporation was that if an 'atom of water' was heated so that it expanded to become a bubble ten times its original diameter, it would became lighter than air, and would consequently rise upwards. With additional heat, more and more particles of water became separated and were emitted with a great velocity, as could be seen in case of boiling cauldrons. The sun heats up the air during the day, and also raises 'more plentiful vapours from the water'. Warm air is capable of holding more

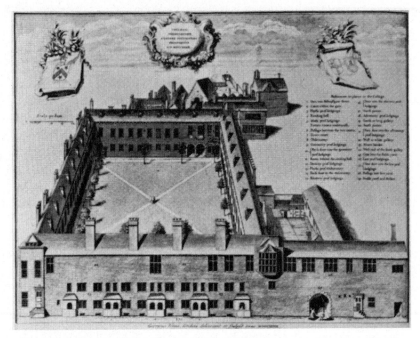

Figure 9. Gresham College, the birthplace and early home of the Royal Society of London, where Halley's evaporation experiments were conducted. From a painting by George Virtue, 1739 (by courtesy of the Society of Antiquaries, London).

aqueous vapour than cool air, and hence during the night when it becomes gradually cooler, some of the vapour is changed to dew. The process is somewhat analogous to the fact that warm water can dissolve more salt than cold water; but when the temperature of the solution drops, some of the dissolved salt is precipitated. Halley suggested that:

'Those vapours therefore that are raised copiously in the sea, and by the winds are carried over the low land to those ridges of mountains, are there compelled by the stream of the air to mount up with it to the tops of the mountains, where the water presently precipitates, gleeting down by the crannies of the stone; and part of the vapour entering into the caverns of the hills, the water thereof gathers as an alembick into the basons of stone it finds, which being once filled, all the overplus of water that comes thither runs over by the lowest place, and breaking out by the sides of the hills, forms single springs.'[25]

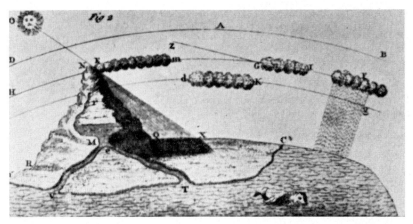

Figure 10. Halley's theory of springs as seen by Switzer.

The springs unite to form rivers which carry the water back again to the sea. Figure 10 is an illustration of Halley's concept of the hydrologic cycle from an early eighteenth century book by Switzer[26] on hydraulics (probably the first English work bearing the title 'hydraulics' – a term introduced by Robert Boyle). Switzer, incidentally, disagreed completely with both Halley's and Mariotte's concepts on these subjects.

Evaporation and the origin of springs
There was a fundamental difference between the explanations of the French scientists and the English astronomer on the origin of springs. Perrault and Mariotte concluded that springs originated from intermittent rains, but Halley stated that water is being continually condensed out of vapour on the long mountain ridges, and: 'it may almost pass for a rule, that the magnitude of a river, or the quantity of water it evacuates is proportionable to the length and height of the ridges from whence its fountains arise'.[25]
In a later paper[27, 28] presented to the Royal Society, Halley demonstrated that enough evaporation takes place from the oceans to more than replenish all springs and rivers. To determine the amount of water evaporating from the oceans, he took a pan of water, 4 in. deep and 7.9 in. in diameter. A thermometer was placed

in the water which was heated to the temperature of the air 'in our hottest summer'. At the end of two hours, he found that 233 grains of water had evaporated. The unit of weight used by the astronomer was the pound troy, now obsolete. The relationship between pound troy and ordinary pound (avoirdupois) is:

1 oz troy = 480 grains = 1.09714 oz avoir, and

12 oz   = 1 lb troy = 0.82286 lb avoir.

Thus the depth of water evaporating from the pan in two hours was

$$\frac{233 \times 76}{1726} \times \frac{1}{49} = \frac{1}{53} \text{ in.}$$

He assumed that 1 cu. ft of water weighs 76 lb troy, and credited Edward Bernard of Oxford for its determination. To simplify subsequent calculations, the depth of evaporation from the pan was taken as 1/120 in. per hour. He considered that the evaporation took place in summer for 12 hours per day, because 'dews return in the night, as much if not more vapours than are then emitted'. Halley calculated that if the Mediterranean Sea was assumed to be 40 degrees (1 degree = 69 miles) long and 4 degrees broad, the total amount of water lost from the sea by evaporation per summer day would be 5,280,000,000 tons. He went on to say that the figure arrived at was very conservative because evaporation depends to a great extent on wind, and its effect was totally neglected in his calculations.

Halley next calculated the amount of water the Mediterranean received from the nine major rivers – Iberus, Rhone, Tiber, Po, Danube, Neister, Borysthenes, Tanais, and Nile. If the river Thames had a cross-sectional area of 300 ft × 9 ft and a mean velocity of 2 mph, the total flow per day would be 20,300,000 tons. Assuming each of the above nine rivers had a discharge equal to ten times that of the Thames, total fresh water received by the Mediterranean per day would be 20.3 × 10 × 9 = 1,827,000,000 tons. Because it is slightly more than one-third the total loss of water, it was proved that enough water evaporated from the ocean to supply all the streams and rivers.

Halley had a reasonably clear conception of the hydrologic cycle:

'Thus is one part of the vapours blown upon the land returned by the rivers into the sea, from whence they came; another part by the cool of the night falls in dews or else in rains; again into the sea before it reaches the land, which is by much the greatest part of the whole vapour, because of the great extent of the ocean, which the motion of the wind does not traverse in a very long space of time. And this is the reason why the rivers do not return so much into the Mediterranean as is extracted in vapour. A third part falls on the lower lands, and is the pabulum of plants, where yet it does not rest, but is again exhaled in vapour by the action of the sun, and is either carried by the winds to the sea to fall in rain or dew there, or else to the mountains to be there to be turned into springs; and though this does not immediately come to pass, yet after several vicissitudes of rising in vapour and falling in rain or dews, each particle of the water is at length returned to the sea from whence it came. Add to this that the rain-waters, after the earth is full sated with moisture, does by the valleys or lower parts of the earth find its way into the rivers, and so is compendiously sent back to the sea. After this manner is the circulation performed . . .'[25]

In the third paper[29] of the series, Halley described the investigation carried out in the Gresham College by Henry Hunt 'with great care and accuracy', under his direction in 1693. The evaporation from a screened and sheltered water surface (having a surface area of 8 sq. in.) was noted every day for the year 1693. Also recorded were temperature (outside?), pressure, and general precipitation conditions (snow, rain, or frost). All observations were taken at 8 a.m. The total annual evaporation was 64 cu. in., or 8 in. of water per sq. in. of the area. He compared it with Perrault's recording of 19 in. of annual rainfall in Paris and Townley's 40 in. at the foot of the hills in Lancashire, but evidently the water evaporated was too little to account for the total annual precipitation. His explanation, for the residual evaporation required to balance the rainfall, was that the direct effects of the sun and wind had been excluded in his experiment. He concluded that the wind effect would have increased the evaporation at least three times, and the sun perhaps might have doubled it.

The experiment also indicated that the evaporation during the months of May, June, July and August are approximately equal, and are about three times the monthly evaporation occurring during November, December, January, and February, and twice as much as that of March, April, September, and October. Regrettably, Halley does not mention anything about the container of water from which evaporation took place. Probably it was a 'pan of water' – like the one used for his previous evaporation experiment, and was filled to the top every morning at 8 o'clock.

The final paper[30] of the series appeared in 1715, and is of considerable interest to all historians of science. Halley considered four closed-in (i.e., having no outlet) seas and lakes – Caspian Sea, Dead Sea, Lake of Mexico, and Lake Titicaca in Peru. He reasoned that as these lakes and seas have no exits, but receive water continuously through various rivers, the levels should rise 'until such time as their surfaces are sufficiently extended, so as to exhale in vapour that water that is poured in by the rivers'. He suggested that as the rivers are continually carrying dissolved salt to the ocean, and the loss through evaporation is only of fresh water, the salinity of the sea must be steadily increasing. Halley concluded that from the degree of salinity, it would be possible to estimate the age of the earth.

CONCLUSION

A study of the development of the science of hydrology will clearly indicate that there have been numerous early contributors to applied hydrology but that their experiments were isolated. Only during the seventeenth century there was a concerted attempt made to establish some fundamental hydrological principles on an experimental basis. The major contributors of that period were Pierre Perrault, Edmé Mariotte, and Edmond Halley. Their contributions to the development of hydrology are truly incalculable. It was Perrault who proved, for the first time by experimental investigations, that rainfall is adequate to sustain stream-flow. His concept of the pluvial origin of springs may have been considerably influenced by Father Jean François (1582–1668) who taught in several Jesuit colleges, and wrote extensively on hydrogeology. His book *La science des eaux qui explique en quatre parties leur formation, communication, mouvements et mélanges*,[31] first published in 1653, was certainly known to Perrault. The opinions of Father François and Perrault on the origin of springs are certainly very similar but unless a thorough study of the Jesuit's works is undertaken it would be impossible to evaluate his influence on Perrault.

Perrault's work was widely known among scientific circles in the seventeenth century, but the posthumous work of Mariotte enjoyed more respect primarily because Mariotte was the more outstanding scientist of his time. Both books, however, received considerable opposition particularly with respect to their concepts of the hydrologic cycle. In fact, with regard to his views on that subject, Perrault could not even convince his brother, Charles, who, along with De la Hire, conducted the following simple experiment in 1690. They buried a clay vessel at a depth of 8 ft and later at 16 ft below the surface and connected it by means of a lead pipe to a cellar. Since no water was discharged from the pipe, it was concluded that rain water could not penetrate more than a few feet of soil. Obviously, the vessel must have been located beneath an impermeable stratum, a most unfortunate circumstance.[32]

The third pioneer, the astronomer Edmond Halley, proved by his calculations that water evaporated from the oceans and came down as rainfall in ample amounts to sustain the flows of rivers. Thus, between the three scientists, the concept of the hydrologic cycle became firmly established even though many noted investigators later refused to believe it. They were the first hydrological investigators to use quantitative results to prove their hypotheses, and that is undoubtedly their greatest contribution to hydrology.

REFERENCES

1. PERRAULT, P., De l'origine des fontaines. Paris, Pierre le Petit (1674).
2. ANONYMOUS, Review of ref. 1, Philosophical Transactions of the Royal Society of London 10 (1675) 447–450.
3. PERRAULT, P., Origin of fountains, translated by A. La Rocque. New York, Hafner Publishing Co. (1967).
4. DELORME, S., Pierre Perrault, auteur d'un traité De l'origine des fontaines et d'une théorie de l'expérimentation. Archives Internationales d'Histoire des Sciences 27 (1948) 388–394.
5. BARBIER, A. S., Dictionnaire des ouvrages anonymes, vol. 3. Paris (1875) p. 746.
6. DOOGE, J. C. I., Quantitative hydrology in the 17th century. La Houille Blanche, no. 6 (1959) pp. 799–807 (in both French and English).
7. MATHER, K. F. and S. L. MASON. A source book in geology. New York, McGraw-Hill Book Co. (1932) pp. 20–23.
8. PERRAULT, P., Ref. 3, pp. 96–98.
9. WATSON, E. C., Edmé Mariotte. The American Physics Teacher 7 (1939) 230–232.

10. MARQUIS DE CONDORCET (M. J. A. N. Caritat), Eloges des académiciens de l'Académie Royale des Sciences morts depuis 1666, jusqu'en 1699. Paris (1733).

11. DARMSTAEDTER, L., The life of Edmé Mariotte. Journal of Chemical Education 4 (1927) 320–322.

12. RÜHLMANN, M., Die technische Mechanik, 2nd ed. Dresden (1845).

13. MAINDRON, E., L'Académie des Sciences. Paris, Felix Alcan (1888) p. 16.

14. MARIOTTE, E., Traité du mouvement des eaux et des autres corps fluides. Paris, E. Michallet (1686).

15. MARIOTTE, E., Review of ref. 14. Philosophical Transactions of the Royal Society of London 16 (1686) 119–123.

16. MARIOTTE, E., A treatise of the motion of water, and other fluids, translation of ref. 14 by J. T. Desaguliers. London (1718).

17. Ibid., p. 16.

18. KEILHACK, C., Lehrbuch der Grundwasser- und Quellenkunde. Berlin, Gebr. Borntraeger (1912).

19. MARIOTTE, E., Ref. 16, p. 189.

20. MARIOTTE, E., Ref. 16, p. 263.

21. ARMITAGE, A., Edmond Halley. London, Thomas Nelson and Sons Ltd. (1966).

22. CHAPMAN, S., Edmond Halley, F. R. S. Notes and Records of the Royal Society of London 12 (1957) 168–174.

23. JONES, H. S., Halley as an astronomer. Notes and Records of the Royal Society of London 12 (1957) 175–192.

24. MCKIE, D. The origins and foundations of the Royal Society of London. Notes and Records of the Royal Society of London 15 (1960) 1–38.

25. HALLEY, E., An account of the circulation of watry vapours of the sea, and of the cause of springs. Philosophical Transactions of the Royal Society of London 16 (1691) 468–473.

26. SWITZER, S., An introduction to a general system of hydrostaticks and hydraulicks, vol. 1. London (1729).

27. HALLEY, E., An estimate of the quantity of vapour raised out of the sea by the warmth of the sun. Philosophical Transactions of the Royal Society of London 16 (1687) 366–370.

28. HALLEY, E., Ref. 27 was also later published in: Miscellanea curiosa, edited by W. Derham, vol. 1. London (1726) pp. 1–12.

29. HALLEY, E., An account of the evaporation of water. Philosophical Transactions of the Royal Society of London 18 (1694) 183–190.

30. HALLEY, E., A short account of the cause of the saltness of the ocean. Philosophical Transactions of the Royal Society of London 29 (1715) 296–300.

31. FRANÇOIS, J., La science des eaux qui explique en quatre parties leur formation, communication, mouvements et mélanges. Rennes, P. Hallaudays (1653).

32. BISWAS, ASIT K., Beginning of quantitative hydrology. Journal of Hydraulics Division, ASCE 94 (1968) 1299–1316.

11

Rain gauges in the seventeenth and eighteenth centuries

INTRODUCTION

Rain gauges, as already have been shown, were used periodically in various parts of the world at different times, viz., in India around the fourth century B.C., Palestine in the first century A.D., China in the thirteenth century, and Korea in the fifteenth century, but they were not used in Europe till about the seventeenth century. The first to have done so on that continent was the Italian, Benedetto Castelli, who made some isolated experiments with a non-recording rain gauge around 1639. There too Sir Christopher Wren devised two of the earliest recording instruments,[1] one of which was later modified by Robert Hooke. But it was not until the latter part of the seventeenth century that a widespread interest began to appear in the construction of various types of rain gauges and in obtaining systematic volumetric measurements of precipitation.

CASTELLI'S LETTER TO GALILEO

It seems probable that Castelli's use of such a gauge led to his having sometimes been erroneously[2, 3] attributed with its invention. The erudite German meteorologist Hellmann[4] had discovered a letter that Castelli had written to Galileo on June 10, 1639, in which he, Castelli, mentioned his use of such a gauge, and Hellmann referred to that discovery in a paper which was published in 1890. In 1891,

Symons stated[2] that probably the earliest measurement of rain was made with it. Because many of the present precipitation measurement techniques were developed by Symons, it is reasonable to assume that later engineers were inclined to accept his contention that Castelli had been the first person to have made such quantitative measurement. Symons had spent much of his life, nearly forty years, in systematizing rainfall data in the British Isles, and had performed careful experiments with regard to the construction, exposure, form and operation of rain gauges.[5] Hellmann, however, later published two papers, one in 1901[6] and the other in 1908,[7] in which he attributed the invention of rain gauges to the Jews in Palestine during the first century A.D., and probably in an effort to correct his previous error, he mentioned that he considered the measurements on which *those* gauges were used to be the earliest quantitative measurements of rainfall.[7] He, incidentally, also mentioned therein that Ferdinand II of Tuscany had a rain gauge in operation in Florence early in the seventeenth century.

In the letter Castelli had written to Galileo Galilei, Chief Philosopher to the Great Duke of Tuscany, he had stated:

'Being returned to Perugia, there followed a rain, not very great but constant and even, which lasteth for the space of 8 hours or thereabouts; and it came into my thoughts to examine, being in Perugia, how much the Lake [Thrasimeno] was increased and raised by this Rain, supposing (as it was probable enough) that the Rain had been universal over all the Lake; and like to that which fell, in Perugia, and to this purpose I took a glasse formed like a cylinder, about a palme high, and half a palme broad; and having put it in water sufficient to cover the bottom of the glasse, I noted diligently the mark of the height of the water in the glasse, and afterwards exposed to the open weather, to receive the raine-water, which fell into it; and I let it stand for the space of an hour; and having observed that in that time the water was risen in the vessel the height of the following line [Castelli here draws a line about 0.4 in. long to represent the depth], I considered that if I had exposed the same rain such other vessel equal to that, the water would have risen in them according to that measure.'[8]

Castelli repeated his experiment and showed it to an engineer having a 'dull brain' (who was not very enthusiastic about the instrument) while it was 'out at my chamber-window exposed in a courtyard'. Symons assumed the rain gauge to be a glass cylinder about 5 in. in diameter and 9 in. deep. Frequently it is contended that Castelli made a rain gauge because of an 'exceptionally heavy

Figure 1. Sir Christopher Wren, from a marble bust by Edward Pierce made about 1673 (by courtesy of the Ashmolean Museum, Oxford).

downpour',[2] but as can be seen from his letter, the rainfall was actually 'not very great, but constant and even'.

That measurement of Castelli's applied only to an isolated rainfall. It does not seem to have occurred to him to use his rain gauge for recording subsequent precipitations. This is borne out by the fact that no references are made to any rain gauges in the Climento manuscripts which date from 1654 to 1664, nor in records of the Monastery of the Angels of Florence of the period 1654 to 1670 where frequent statements can be found concerning the dates on which rain or snow fell. It is reasonable to assume that if a rain gauge had actually been in operation, details thereon as well as the results of any measurements would have appeared in those manuscripts.

SIR CHRISTOPHER WREN'S FIRST RAIN GAUGE

The earliest English rain gauge[9-13] was made by Sir Christopher Wren (1632–1723; figure 1) and, unlike the previous instruments, this one was of a recording type. There is no evidence, however, of it ever having been used for obtaining regular observations of rainfall.

B. De Monconys, a Frenchman, who visited England in June 1663, described the automatic rain gauge in question. He reported it as having been made by one 'M. Renes'.[14] It is contended here that 'M. Renes' was actually Sir Christopher Wren, and that the poor Frenchman failed to realize that the surname, pronounced 'Ren', was actually spelled Wren. Such a mistake is quite understandable, and is, in fact, believed to have occurred in this instance. The rain gauge which De Monconys described is shown in figure 2. It was, as will be seen, a part of a meteorograph. Below the catch funnel was a three-compartment container mounted on a rack which was moved slowly forward by clockwork in such a manner that one of those compartments would collect any rain which would fall during the first hour; the next compartment would collect any which fell during the second hour, etc.

Middleton[15] has suggested that assuming De Monconys' diagram has some validity as an elementary sketch, it was probably a rough model constructed by Wren to tidy up his initial ideas. Probably Wren was reluctant to show the weather-clock to the Royal Society

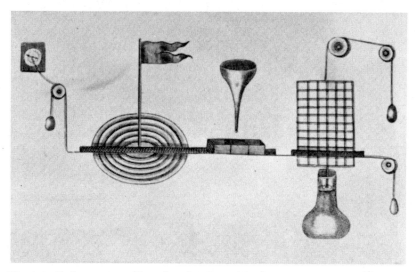

Figure 2. Rain gauge attributed to 'M. Renes' by De Monconys.

in its underdeveloped form. Middleton further suggested that Wren later substituted rack and pinion arrangement for the drum and cord of the first meteorograph.

THE WREN–HOOKE TIPPING-BUCKET RAIN GAUGE

Sir Christopher Wren's second automatic rain gauge was of the tipping-bucket type, provided with recording facilities. It was a part of an all-purposes 'weather-wiser' discussed as follows by Nehemiah Grew (1641–1712), a former secretary of the Royal Society:

'Begun by Sir Christopher Wren, now President of the Royal Society. To which other motions have since been added by *Mr. Robert Hooke* Professor of Geometry in *Gresham-Colledge*. Who purposes to publish a description hereof. I shall therefore only take notice, that it hath six or seven motions; which he supposeth to be here advantagiously made altogether. First a *pendulum* clock, which goes with $^3/_4$ of a 100 *Lib*. weight, and moves the greatest part of the work. With this, a barometre, a thermometre; a rain-measure, such an one as is next describ'd; a *weather-clock*, to which subserves a piece of wheel-work analogous to a *way-wiser*; and a hygroscope. Each of which have their register, and the *weather-clock* hath two, one for the *points*, the other for the strength of the wind. All working upon a paper falling off a rowler which the *clock* also turns.

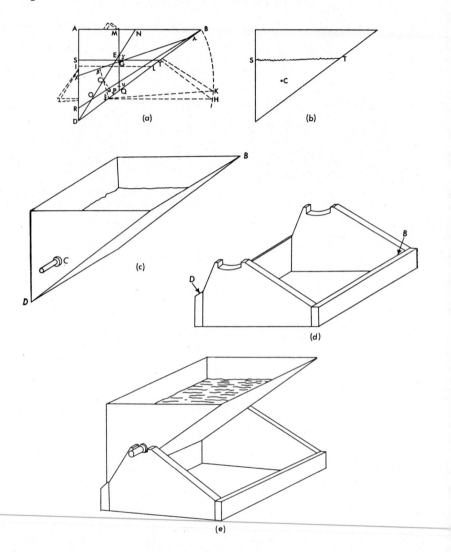

Figure 3. (a) The original sketch of Robert Hooke. (b) All construction lines have been moved. Point C is the centre of gravity of the prism of water contained in the vessel when it has been filled to the elevation of S–T (the incipient point above which tipping will occur). (c) This shows the vessel in three dimensions. (d) Hooke mentions that the vessel is 'poiz'd like a balance upon a foot'. Maybe the foot looked something like this. Surfaces D and B serve as the stops to limit the travel of the vessel. (e) When assembled, the tipping-bucket rain gauge might have looked like this (by courtesy of Arthur H. Frazier).

An instrument for MEASURING the quantity of RAINS that fall in any space of time, on any one piece of ground, as suppose upon one acre in one year. Contrived by *Sir Christopher Wren*. In order to the *theory* of vapours, river, seas, &C. A triangular tin-vessel hanging in a frame, as a bell, with one angle lowermost. From whence one side rises up perpendicular, the other sloaped; whereby the water, as it fills, spreads only on one side from the centre, till at length it fills and empties itself. Which being one, a leaden poise, on the other side, immediately pulls it back to fill again.'[16]

Hooke both described the gauge and furnished a drawing explaining its construction. Figure 3a shows his original drawing thereof as first reproduced by Derham[17] (in 1726), and later (1930) by Gunther.[18] In figure 3b, point C indicates the centre of gravity of the prism of water contained in the vessel when it has been filled to the elevation of S–T (the incipient point above which tipping will occur). Hooke mentions that the vessel is 'poiz'd like a balance upon a foot'. Perhaps the foot looked something like that in figure 3d, in which the surfaces D and B serve as stops for limiting the number of degrees through which the vessel could tilt. When finished, Hooke's tipping-bucket rain gauge may have looked like the assembly illustrated in figure 3e.

Two alternative methods for counterpoising the weight of the bucket were contrived by Hooke. The first used a string of bullets so arranged that when the bucket was empty, all of the bullets would lie on a table. They would be lifted up one after another as more and more water accumulated in the bucket. By the time all of them were lifted from the table, the bucket would empty itself. Hooke rejected that arrangement on the ground that the movement of the bucket would neither be smooth nor continuously equal.

The second method was described as using a

'counterpoise to the bucket, when empty was a cylinder immersed into water, mercury or any other fluid; which cylindrical counterpoise, according as the bucket received more and more water, was continually lifted higher and higher out of the water by spaces, always proportioned to the quantity of water, that was contained in the bucket. And when the bucket was filled to its designed fullness, it immediately emptied itself of the water, and the cylinder plunged itself into the water, and raised the bucket to the place where it was again to begin its descent.'[17, 19, 20]

The principle of the tipping-bucket rain gauge was certainly known before Wren's time. For example, Muhammad ibn Ibrāhīm, al-

Figure 4. The principle of 'tipping-bucket' type of rain gauge was known to the Arabs. The illustration (from a manuscript on *Automata* by Muhammad ibn Ibrāhīm, al-Jazarī, dated 1364) shows an automatum in which two figures pour wine for each other. The automatum is actuated by wine stored in the dome; through the action of the 'tipping bucket' above the figures, the wine flows first to one man and then to the other (by courtesy of the Museum of Fine Arts, Boston).

Jazarī, in his treatise on automata (c. 1364) describes an automatum in which two figures alternately pour wine for each other (figure 4). It was actuated by wine stored in the dome thereof. Through the action of the tipping-bucket above the two figures, the wine would flow first to one of them and then to the other. Although the principle was known, up to now there is no evidence to indicate that any tipping-bucket type *rain gauge* had been built before the time of Sir Christopher Wren.[21]

The weather-clock, of which the Wren–Hooke self-measuring rain gauge was a part, consisted of two parts. The first part consisted of a strong and large pendulum-clock designed to measure time as well as to unwind a paper strip into which the records would be punched every 15 minutes. The second part consisted of an arrangement of five meteorological devices: a barometer, thermometer, hygroscope, rain bucket, and finally a revolving type wind vane, all of which provided the data that were punched through the paper strip. A further description of it follows:

'The stations or places of the first four punches are marked on a scrowl of paper, by the clock-hammer, falling every quarter of an hour. The punches, belonging to the fifth, are marked on the said scrowl, by the revolutions of the vane, which are accounted by a small numerator, standing at the top of the clock-case, which is moved by the vane-mill.'[17, 22]

The punched records of the observations from the rain-bucket not only showed the number of tippings as they occurred, but also indicated the quantity of water that remained in it.

The meteorograph obviously worked, because Hooke and his assistant were asked by the Royal Society to 'reduce into writing' some of the data from the punched tapes. The fact that the punched tape method of registering data has come into vogue only during the last decade or two, clearly indicates that Hooke, who used it as early as 1678, was well ahead of his time. To the author's knowledge, no sketch or specimen of the actual punch tape used exists at present, but it may be noted in this connection that Jacob Leupold in his book *Theatri machinerum supplementum*, published in 1739, described with sketches[23] the use of a similar punched tape to record results obtained from way-wisers (pedometers).

The chronology of Wren's recording rain gauge is described[24] by

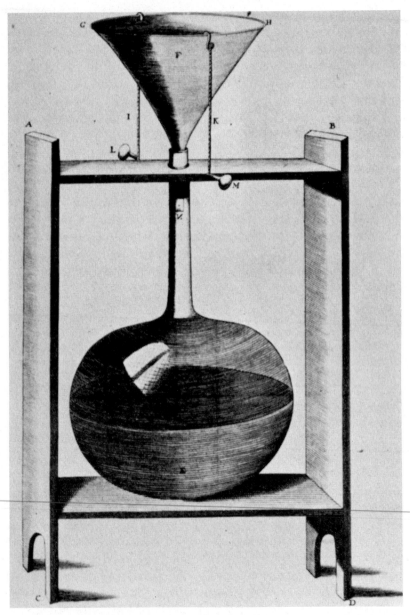

Figure 5. Non-recording rain gauge of Robert Hooke.

Thomas Birch (1705–1766), the then secretary of the Royal Society, and has been discussed in considerable detail by Biswas.[1]

Hooke also contrived a non-recording type of rain gauge[25] as shown in figure 5. It was used in 1695 at Gresham College where Hooke was a professor of geometry. A large flask capable of holding more than 2 gallons, called a 'large bolt head', was supported by a wooden frame having, presumably, a 11.4-in.-diameter glass catch funnel. The funnel was held steady against the wind by two stays or pack threads strained by two pins. The neck of the container was 20 in. long by 0.2 in. in diameter to minimize evaporation. The rain-

Months	Days	lb	℥	Gr.	Months	Days	lb	℥	Gr.
August	19	2	6	215	March	2	0	9	12
	26	4	6	246		9	0	2	459
September	2	9	4	96		16	0	0	396
	9	3	10	397		23	4	4	263
	16	0	1	204		30	1	5	285
	23	0	6	336	April	6	2	3	375
	30	4	1	444		13	1	0	294
October	7	2	3	96		20	2	1	000
	14	0	2	60		27	0	7	390
	21	0	1	234	May	4	4	10	45
	28	0	0	45		11	7	6	000
November	4	0	0	207		18	6	2	105
	11	1	11	65		25	1	7	60
	18	1	1	309	June	1	0	0	99
	25	0	9	285		8	6	6	150
December	2	0	8	126		15	0	2	120
	9	3	7	324		22	7	5	285
	16	1	3	435		29	1	5	84
	23	0	1	60	July	6	0	1	120
	30	5	8	93		13	16	1	000
January	6	4	10	105		20	1	7	240
	13	0	1	12		27	6	1	256
	20	1	10	450	August	3	1	10	120
	27	1	5	82		10	1	11	90
February	3	5	11	372		12	0	0	0
	10	4	9	242	The Sum		131	7	113
	17	0	6	291					
	24	0	2	180					

$=$ to $29\frac{11}{12}$ Inches in a Cylinder of the aforesaid Diameter, viz. $11\frac{3}{2}$ Inches.

Figure 6. Typical record of observations of Hooke's rain gauge.

water which accumulated in the container was measured every Monday, and the total precipitation over a period of time was expressed as a vertical depth which had fallen during that time interval (see figure 6).

OBSERVATIONS BY TOWNLEY

The first continuous observations of rainfall in Britain were made by Richard Townley (1629–1707) of Townley Hall in Lancashire,[26] starting from 1677. For his rain gauge he

'... fixed a round tunnel of 12 inches diameter to a leaden pipe, which could admit of no water, but what came through the tunnel, by reason of a part soder'd to the tunnel itself, which went over the pipe, and served also to fix it to it, as well as to keep out any wet that in stormy weather might beat against the under part of the tunnel, which was so placed, that there was no building near it that would give occasion to suspect that it did not receive its due proportion of rain that fell through the pipe some nine yards perpendicularly, and then was bent into a window near my chamber, under which convenient vessels were placed to receive what fell into the tunnel, which I measured by a cylindrical glass at a certain mark, containing just a pound, or 12 ounces troy, and had smaller parts also.'[27]

The tunnel was fixed on the roof of Townley's house. Symons[2] later conducted some experiments with a similar gauge having 27 ft of pipe, and concluded that its evaporation loss was almost imperceptible.

For the first 6 months of 1699, Derham[28] measured the rainfall at Upminster. No details are available of that rain gauge, but it was probably similar to Townley's.

MEASUREMENTS BY PERRAULT AND MARIOTTE

Very little is known[29, 30] about Pierre Perrault's rain gauge (see chapter 10). His compatriot Edmé Mariotte also conducted[31] a comparison of rainfall and runoff on the river Seine catchment above Paris, and it was an improvement over Perrault's. Rainfall measurements were made at Mariotte's request by 'a very skillful man and very exact in his experiments', probably around 1678.

'He placed near the top of the house a square vessel, about two foot diameter, at the bottom of which there was a pipe which conveyed the rain that fell into it into a cylindric vessel, where it was easy to measure it as often it rained; for when the water was in the cylindric vessel, there was very little exhaled during five or six days. The vessel of two foot diameter was sustained by a bar of iron, which advanced about six foot beyond the window, whereon it was placed and fixed that it might receive only rain-water, which fell immediately upon the breadth of its opening, and that there might not enter any but what was to fall according to the proportion of upper surface.'[32]

EARLY EIGHTEENTH CENTURY

The interest in the measurement of precipitation increased considerably during the earlier part of the eighteenth century throughout the world. According to Horton, the physicians Kindmann and Kanold of Breslau, Prussia, invented conical rain gauges[33] around 1717, and they obtained measurements with them throughout the period 1717 to 1727. Probably the earliest approximation of the modern non-recording instruments were made by Horsley in England in 1722. Horsley realized that

'. . . weighing the water and reducing it from weight to depth seemed pretty troublesome, even when done in the easiest method: to remedy this inconvenience (besides a funnel and proper receptacle for the rain) I use a cylindrical measure exactly 3 inches, the depth of the measure is 10 inches, and the gauge of the same length with each inch divided into 10 equal parts; or, instead of a gauge, the inches and divisions may be marked on the side of the cylindrical measure. The apparatus is simple and plain, and it is easy to apprehend the design and reason for the contrivance; for the diameter of the cylindrical measure being just $1/_{10}$ of that of the funnel, and the measure exactly 10 inches deep, 'tis plain that 10 measures of rain make an inch in depth. . . By this means the depth of any particular quantity which falls, may be set down with ease and exactness and the whole at the end of each month or year may be summed up without trouble.'[34]

In 1723, James Jurin drew up a set of rules for providing uniformity in meteorological observations.[35]
Many different types of rain gauges were used after 1710. Some important ones – including a few unusual examples – will be briefly described below.

Rain gauges of Leupold
Leutinger[36] described a hyetometer in 1725 which, according to Leupold,[37] was first devised by Leutmann. It had a square funnel

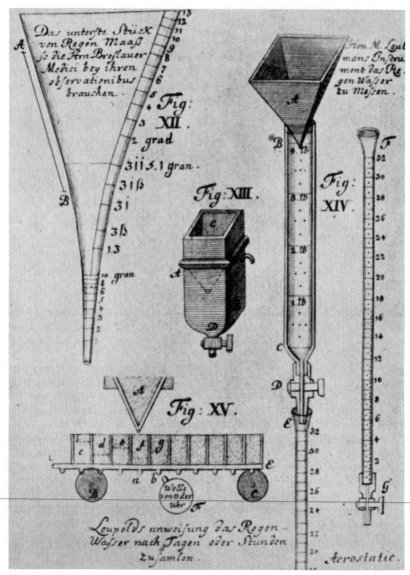

Figure 7. Non-recording rain gauges described by Leupold.

leading to two glass tubes (figure 7, Fig. XIV) and measured the total weight rather than the depth of rainfall. The two tubes were calibrated to give the weight of rain water collected. There were two taps in the instrument. No records are available to indicate whether this instrument was actually put into practical use.

Leupold also described various other types of rain gauges, for example, the one (figure 7, Fig. XII) used by the Breslau Natural History Society started in 1717 (the same one mentioned by Horton?) which had a sharp-edged glass funnel of about 4 in. in diameter and a depth of 8 in. As with the Leutmann's gauge, this particular type indicated the total weight of accumulated rainfall rather than its depth.

Leupold's own non-recording type instrument (figure 7, Fig. XIII) had a 9-in. square receiver. The rainfall was collected in a glass tube for determining its quantity. Two automatic gauges of his have also been described. The first had a series of compartments beneath a catch funnel, and the compartments were moved by clockwork so that the funnel remained above each of those receptacles for a measured period of time (figure 7, Fig. XV). If the quantity of water was measured in each compartment, it would represent the total amount of rain that had fallen during the particular time interval involved. The second of those two automatic rain gauges – 'Leupold's hyetometer' – was of a tipping-bucket type in which the rainfall collected by a funnel having a square opening was directed into a small bucket at the end of a balance level. Each time the bucket became filled with rainwater it tipped and emptied itself. While so doing, the first wheel of a counting device was advanced by one tooth (figure 8). As there were four wheels, the gauge could automatically indicate as many as 10,000 tips. While tipping occurred, a special device prevented any water from by-passing the measuring bucket. The most striking difference between the English and the continental gauges of this period was in the use made of taps for emptying the measuring flasks. The continentals used them extensively, whereas the English seldom used them because they realized that taps tended to leak after short periods of use.

Rain gauges of Pickering, Dobson, and Dalton
In 1744, Pickering proposed a rain gauge having a tin funnel of

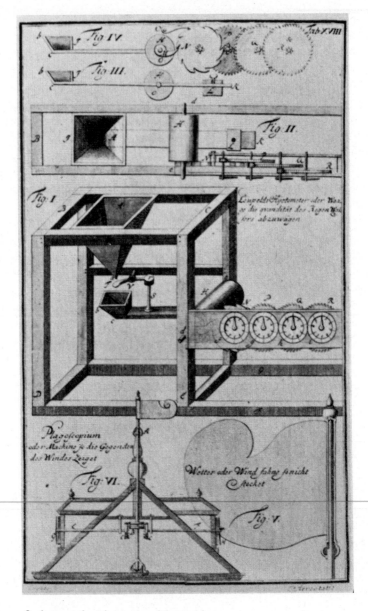

Figure 8. Automatic rain gauge of Leupold.

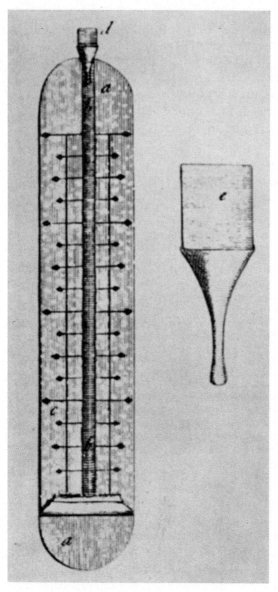

Figure 9. The rain gauge of Pickering.

1 sq. in. area (figure 9) which discharged into a $^1/_2$-in.-diameter glass tube that was more than 2 ft long. The entire instrument was mounted on a 3-ft-long board which was hung against the rail at the top of his house.[38] The inch graduations on the tube were divided into 32 parts. He claimed that with a diameter of tube that small, the results would be more accurate than those obtained with instruments using larger tubes. The instrument was very simple and could be repaired easily if the tube became broken.

Dobson was among the first (as of 1777) to measure rainfall in a manner that corresponded with present standards. Most rain gauges used during this period (and even later) were placed on the roofs of houses so that they might 'record free fall of rain'. His rain gauge was a well-varnished 12-in.-diameter tin funnel which was fixed on to the top of a large stone bottle by means of a grooved cork, the use of which was primarily intended to reduce evaporation. He chose a site at the middle of a grassy plot overlooking Liverpool, 75 ft above sea level, on rising ground having free exposure to the sun, wind, and rain.[39] His primary interest seems to have been concerned with evaporation.

Dalton's definition of rain gauge and its description was brief and precise:

'The rain gauge is a vessel placed to receive the falling rain, with a view to ascertain the exact quantity that falls upon a given horizontal surface at the place. A strong funnel, made of sheet iron, tinned and painted, with a perpendicular rim two or three inches high, fixed horizontally in a convenient frame with a bottle under it to receive the rain, is all the instrument required.'[40]

CONCLUSION

No hydrological or meteorological instrument has received attention so constantly and for such a long period as the rain gauge. The interest in mean annual rainfall, however, made its first appearance around the middle of the eighteenth century. The prevailing attitude of that period is reflected by the statement of Gilbert White[41] who said that his own observations (May, 1779 to December, 1786) were too short to provide a good estimate of the mean rainfall. Credit for keeping the longest rainfall record (by a single person using the same instrument) during the seventeenth and eighteenth centuries must go to White's brother-in-law, Thomas Barker of Lyndon in Rutland, England, who did so for a period of 59 years (1736 to 1796).

A major defect appears frequently in one of the average annual rainfall records published for the eighteenth century. Those responsible for it have failed to note that September, 1752, had officially only 19 days.

The present chapter can be best concluded by quoting G. J. Symons, to whom hydrologists and meteorologists all over the world owe their gratitude for his rationalization of rainfall data:

'... I desire to meet at the very outset an objection sometimes raised, viz., that we cannot trust very old observations ... I maintain that we can trust them, ... I think them far more reliable than many modern ones; for in the 17th and early part of the 18th century, to measure the fall of rain was esteemed a serious undertaking, only to be accomplished by first-class men.'[42]

REFERENCES

1. BISWAS, ASIT K., The automatic rain gauge of Sir Christopher Wren. Notes and Records of the Royal Society of London 22 (1967) 94–104

2. SYMONS, G. J., A contribution to the history of rain gauges. Quarterly Journal of the Royal Meteorological Society 17 (1891) 127–142.

3. METEOROLOGICAL OFFICE, AIR MINISTRY, Handbook of meteorological instruments. London, Her Majesty's Stationary Office (1956) p. 258.

4. HELLMANN, G., Die Anfänge der meteorologischen Beobachtungen und Instrumente. Berlin, Druck von Wilhelm Gronau (1890) pp. 1–24. Also published in Himmel und Erde, vols. 2–4 (1890).

5. KURTYKA, J. C., Precipitation measurements study. Report of Investigation no. 20, State Water Supply Division, Urbana, Ill. (1953).

6. HELLMANN, G., Die Entwickelung der meteorologischen Beobachtungen bis zum Ende XVII Jahrhunderts. Meteorologische Zeitschrift (1901) 145–157.

7. HELLMANN, G., The dawn of meteorology. Quarterly Journal of the Royal Meteorological Society 34 (1908) 221–232.

8. CASTELLI, B., Of the mensuration of running water, translated by T. Salisbury. London, William Leybourn (1661) p. 28.

9. BISWAS, ASIT K., Development of rain gauges. Journal of Irrigation and Drainage Division, ASCE 93 (1967) 99–124.

10. BENTLEY, R., The growth of instrumental meteorology. Quarterly Journal of the Royal Meteorological Society 31 (1905) 182–185.

11. WOLF, A., History of science, technology, and philosophy in the 16th and 17th centuries. London, George Allen and Unwin Ltd. (1935) p. 310.

12. MIDDLETON, W. E. K., Invention of meteorological instruments. Baltimore, Johns Hopkins Press (1968).
13. MULTHAUF, R. P., The introduction of self-registering meteorological instruments. Bulletin 228, US National Museum, US Government Printing Office, Washington, D.C. (1961).
14. DE MONCONYS, B., Journal des voyages de Monsieur de Monconys, 1st part. Lyons, Chez Horace Boissat et George Remeus (1666) pp. 54–55.
15. MIDDLETON, W. E. K., The first meteorographs. Physic, Rivista di Storia della Scienza 3 (1961) 213–222.
16. GREW, N., Musaeum Regalis Societatis. London (1681) pp. 357–358.
17. DERHAM, W., Philosophical experiments and observations of the late eminent Robert Hooke. London, W. and J. Innys (1726) pp. 41–47.
18. GUNTHER, R. T., Early science in Oxford, vol. VII. Oxford, University Press (1930) pp. 520–522.
19. Ibid., pp. 515–516.
20. BIRCH, T., The history of the Royal Society of London, vol. 3. London (1757) pp. 476–477.
21. IBRAHIM. MUHAMMAD IBN (AL-JAZARI), Treatise of al-Jazari on automata. Manuscript. Museum of Fine Arts, Boston (1364).
22. GUNTHER, R. T., op. cit., pp. 519–520.
23. LEUPOLD, J., Theatri machinarum supplementum. Leipzig, B.C. Breitkopf (1739) p. 20.
24. BIRCH, T., The history of the Royal Society of London, 4 vols. London (1757).
25. HOOKE, R., An account of the quantities of rain fallen in one year in Gresham College, London, begun August 12, 1695. Philosophical Transactions of the Royal Society of London 19 (1695) 357.
26. WEBSTER, C., Richard Townley, 1629–1707, and the Townley Group. Transactions of the Historic Society of Lancashire and Cheshire 118 (1966) 51–76.
27. TOWNLEY, R., A letter from Richard Townley Esq., of Townley in Lancashire, containing observations on the quantity of rain falling monthly for several years successively. Philosophical Transactions of the Royal Society of London 18 (1694) 52.
28. DERHAM, W., Part of a letter from the Rev. Mr. Derham to Dr. Sloane, giving an account of his observations of the weather for the year 1699. Philosophical Transactions of the Royal Society of London 22 (1700) 527–529.
29. ANONYMOUS, Early measurements of rainfall and streamflow. Engineering News-Record 86 (1921) 379.
30. PERRAULT, P., De l'origine des fontaines. Paris, Pierre le Petit (1674).
31. MARIOTTE, E., Traité de mouvement des eaux et des autres corps fluides. Paris, E. Michallet (1686).
32. MARIOTTE, E., A treatise of the motion of water and other fluids with the origin of springs and cause of winds, translation of ref. 30 by J. T. Desaguliers. London (1718).

33. HORTON, R. E., The measurement of rainfall and snow. Journal of the New England Water Works Association *33* (1919) 14–23.

34. HORSLEY, REV. S., An account of the depth of rain fallen from April 1, 1722, to April 1, 1723. Observed at Widdrington in Northumberland. Philosophical Transactions of the Royal Society of London *32* (1723) 328–329.

35. JURIN, J., Invitatio ad observationes meteorologicas communi consilio instituendas. (An invitation to an association for forming meteorological diaries with a specimen.) Philosophical Transactions of the Royal Society of London *32* (1723) 422–427.

36. LEUTINGER, N., Instrumenta meteorognosiae inservientia. Wittenbergae (1725).

37. LEUPOLD, J., Theatri statici universalis. Leipzig, gedruckt von C. Zunkel (1726).

38. PICKERING, R., A scheme of a diary of the weather; together with draughts and descriptions of machines subservient thereunto. Philosophical Transactions, Royal Society of London *43* (1744) 1–18.

39. DOBSON, D., Observations of annual evaporation at Liverpool in Lancashire. Philosophical Transactions of the Royal Society of London *67* (1777) 244–259.

40. DALTON, J., Experiments and observations to determine whether the quantity of rain and dew is equal to the quantity of water carried off by the rivers and raised by evaporation; with an enquiry into the origin of springs. Memoirs of the Literary and Philosophical Society of Manchester *5* (1802) 346–372.

41. WHITE, G., The natural history of Selbourne. Bohn's Illustrated Library. London, G. Bell and Sons (1849).

42. SYMONS, G. J., On the rainfall of the British Isles. Report of the 35th Meeting of the British Association for the Advancement of Science, Birmingham, Sept. 1865. London, John Murray (1866) pp. 192–242.

12

The eighteenth century

INTRODUCTION

Shortly before his death, in 1727, Isaac Newton said:

'I do not know what I may appear to the world, but to myself I seem to have been only like a boy playing on the sea shore, and diverting myself in now and then finding a smoother pebble or a prettier shell than ordinary, while the great ocean of truth lay all undiscovered before me.'

Newton's observation was very valid so far as hydrology is concerned. The development of the subject, even at the beginning of the eighteenth century, was extremely modest, and very few fundamental principles have been realized, let alone universally accepted. The establishment of the learned societies in England, France, and Italy during the latter half of the seventeenth century provided the impetus necessary for the rapid development of natural and physical sciences. The downfall of the old masters was rapidly nearing completion. The motto of the Royal Society of London was *Nullius in verba* (on the words of no man). According to Cardwell:

'Although the scientific achievements of the 18th century were substantial, the technological triumphs were of at least equal interest. The men associated with these triumphs, men like Newcomen, Smeaton, Watt, Wedgewood, etc., were scientific technologists, capable of using scientific method and knowledge in their practical work and often, in return, making contribution to 'pure' science. The rise of these scientific engineers was, paradoxically unaccompanied by a systematic development of applied science.'[1]

It was the same with hydrology.

ANTONIO VALLISNIERI AND MARQUIS POLENÌ

One of the leading figures of hydrology of this period was Antonio Vallisnieri (1661–1730), President of the University of Padua, who published a treatise[2] in 1715 on the origin of rivers, based mainly on his personal observations in the Alps. He was aware of the works of Perrault and Mariotte. He observed the mountain ranges from which many of the Italian rivers originated in order to verify the pluvial theory. He found no indication of sea-water being forced out of the mountain tops, instead he found water always trickling down the slopes. He saw the continuous presence of snow and ice high up in the mountain ranges, and reasoned that its melting along with rainfall, provided the necessary supply of water to all springs and rivers.

He was surprised to find that very few small springs originated from the extensive snowfields of the Pellegrino Alps. The local shepherds, however, showed him the reason for the anomaly. The water from the constant reservoir of melting snow travelled downward, towards Modena, through hidden subterranean channels. He quoted Dante to express his feelings after the discovery of that phenomenon: 'Like a man who when in doubt is reassured, and whose fear changes into comfort because the truth was now revealed to him.'[3] He was quick to realize this was the source of the artesian wells of Modena about which there had been considerable speculations in the past. The reason was quite simple. The subterranean streams, originating from the Pellegrino Alps, passed below Modena towards Bologna. They obviously flowed under great pressure, and when a well was sunk in Modena (see Ramazzini's description in chapter 9), water gushed up to the surface and formed the artesian wells. These explanations of the source of artesian water and their mechanism were even better than those of Ramazzini.

Vallisnieri's book was illustrated by six geological sections (figure 1) which showed the structures of certain mountain ranges in Germany and Switzerland. They were drawn by the naturalist Scheuchzer, who explored the Alps between 1702 and 1711, and according to Adams, they were among the earliest geological sections ever drawn. But the Italian's concepts did not go unchallenged. Gualtieri (and some others) violently disagreed with this heretic who dared to

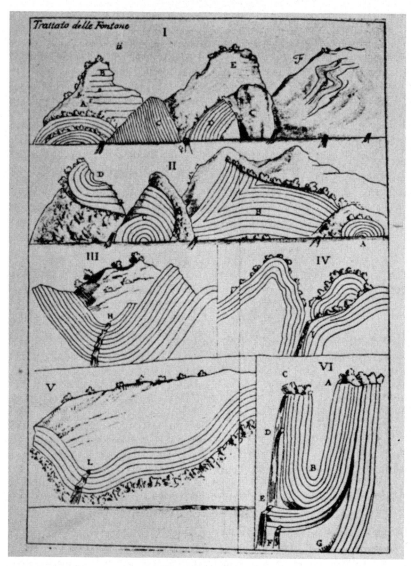

Figure 1. Geological sections of certain mountains in Switzerland and Germany as drawn by Scheuchzer.

dispute the Holy Writ. In the second edition of Vallisnieri's book were discussions of his work by various Italian authors.

Another Italian of note during the early eighteenth century period was Marquis Giovanni Poleni (1683–1761). Born in Venice, he had the distinction of having become a professor of astronomy at the early age of twenty-six at the University of Padua. Later he became a professor of physics and finally a professor of mathematics at the same university. He also served as a consultant in the field of flood control and water-supply engineering. In his treatise,[4] published in 1717, he analysed the flow of water through a rectangular opening which extended to the free surface. Assuming a parabolic velocity distribution curve, he obtained the rate of discharge per unit width as:

$$Q = \tfrac{2}{3}hb\sqrt{f}$$

where, h and b = depth and breadth of opening, and f = velocity function $(2gh)$.

Later the analogy was extended to flow over sharp crested weirs. The resulting equation,

$$Q = \tfrac{2}{3}Cb\sqrt{2g}\,h^{3/2},$$

was named after Poleni.

CAPILLARY THEORY OF SPRINGS

Reverend W. Derham (1657–1735), in his book *Physico-theology*, first published in 1713,[5] put forward the capillary theory of the origin of springs. Switzer described the concept as follows:

'As to the manner how waters are raised up into mountains, and other high lands, and which has all along puzzled so many great men (Mr. Derham says) may be conceived by an easy and natural representation, made by putting a little heap of sand or ashes, or a little loaf of bread, into a basin of water, where the sand will represent the dry land, or an island, and the bason of water the sea about it; and as the water in the bason rises up to or near the tops of the heap in it, so does the water of the sea, lakes, etc. rise in hills: which case he takes to be the same with the rise of liquids in capillary tubes, or between contiguous plains or in a tube fill'd with ashes . . .'[6]

Switzer fully agreed with Derham's concept and opined that the origin of streams and rivers cannot be entirely due to precipitation. He contended that as Derham's idea was based on his own meteorological observations (see chapter 11), they were very exact, and hence, beyond any dispute. Had either of them made a simple experiment to determine the maximum height of capillary rise, their opinions would have changed soon enough.[7]

One of the major chapters of the book *Le spectacle de la nature* by N. A. Pluche (1688–1761), published in 1732 ,was devoted entirely to the origin of springs. The capillary theory was put forward very clearly and concisely by one of the characters of the book:

'I firmly believe that the sea-water deposits its salt on the sands below, and that it rises by little and little, distilling through the sands, and the pores of the earth, which have such a power of attraction as is not easily accounted for; and that not only sand, but other earthy bodies have the power of attracting water, I am well assur'd of from an observation which occur'd to me but this very day. When I threw a lump of sugar into a small dish of coffee, I found that the water immediately ascended thro' the sugar, and lay upon the surface of it. Yesterday I observed, likewise, that some water which had been pour'd at the bottom of a heap of sand, ascended to the middle of it. And the case, as I take it, is exactly the same with respect to the sea and the mountains.'[7]

The main character, the pundit of the book, vigorously opposed the theory on three counts. Firstly, water cannot rise more than 32 ft in dry sand, and even then that height is very seldom achieved. Secondly, the growth of algae will prevent the passage of water after some time, and finally, if it was true, sea-water, due to the same reason, would saturate all the plains adjoining the coast. Pluche believed in the pluvial origin of springs, and firmly discounted Descartes' concept on the subject. He calculated that if a cu. ft of sea-water contained only one pound of salt instead of the usual two, the daily flow of the river Seine alone (288 million cu. ft, as calculated by Mariotte) will deposit 288 million pounds of salt every day. Obviously, the quantity of salt that would be deposited by all the rivers of the world would be too vast for the theory to be true.

VELOCITY DETERMINATION BY THE PITOT TUBE

Henry De Pitot (1695–1771), born in Aramon in south-western
France, was a student of the well-known scientist Réaumur at Paris.
He became the superintendent of the Canal du Midi in his native
province of Languedoc, in 1740, and was concerned with the construction of various flood control works, bridges, and aqueducts, and
drainage of marsh lands. His greatest claim to fame, however, rests
primarily on the invention of a very simple device, presently known
as the Pitot tube.

De Pitot, in his paper[8,9] of 1732, discussed the importance of
velocity distribution in rivers, and also reviewed the existing state
of knowledge on the subject. He outlined the two theories on the
variation of velocities with depths, and preferred the concept that
velocity at the bottom of a river would be less than at the top
because of the frictional resistance. He did not favour the use of
floats to estimate velocities as the method was inaccurate on several
counts. Firstly, a wax sphere is not always visible, and if a piece of
wood, large enough to be kept in sight, is used as a float, it would
encounter air currents which were likely to introduce errors.
Secondly, repeated experiments conducted within the same stretch
of a river would give different results as floats do not travel along a
fixed course. Thirdly, it is almost impossible to measure correctly
the distance travelled by a float during a certain time. Finally, it
only measures surface velocities, and thus, velocities at different
depths cannot be determined by this method. All these problems
could be surmounted easily with his new instrument which had the
added advantage that its operation was as simple as 'plunging a
stick into water'.

De Pitot's 'machine' consisted of two parallel tubes, one straight
and the other bent through 90° for a short length at the lower end
(figure 2), mounted on a wooden frame having a scale. The instrument
is immersed in water to the desired depth with the bent tube facing
into the current. In still water, the levels of water in both tubes
will be identical but in flowing water a difference of elevations
would occur, the extent of which depended on the velocity.

De Pitot was excited about his new instrument:

'The idea of this machine is so simple and so natural, that the moment it occurred to me, I ran immediately to the river bank to make a first experiment with a simple glass tube, and the result confirmed completely my conviction. After this first experiment, I could not imagine how so simple and useful thing could have escaped so many skilled people who have written and experimented on the motion of water.'[10]

The new instrument, conceived by theoretical analysis, was an extraordinary case wherein two erroneous concepts were used to arrive at a correct solution. De Pitot's explanations of efflux and resistance laws were as follows:

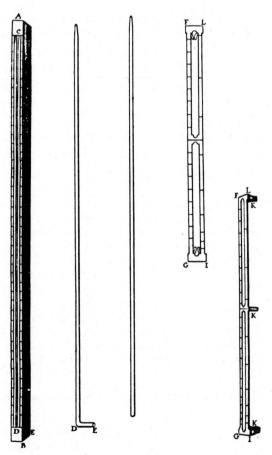

Figure 2. De Pitot's drawings of his tube.

'There is no one with a slight knowledge of the theory of the motion of water who will not conceive immediately the effect of this machine; because, according to the first principles of this science, one must consider the velocity of flowing waters as a velocity obtained from a fall of a certain height, and that if the water moves upward with an acquired velocity, it will rise exactly to the same height. . .'

'Furthermore, the force of impulse of water due to its velocity is equal to the weight of column of water, which has for the base the area of the surface on which the water is impinging and for height that from which the water must have fallen in order to acquire that velocity. Thus the water must rise in the tube of our machine due to the force of a current to exactly the same height from which it must have fallen to form this current.'[8]

It should be remembered that because the value for g had not yet been evaluated, the equation $v = \sqrt{2gh}$ was not correctly proved during De Pitot's time, and Varignon, whom De Pitot credited as being the first to 'have the glory of demonstrating this principle', had not advanced beyond the form $v = \sqrt{gh}$. The impulse–momentum relationship, as used by De Pitot, was Newton's improvement of the postulate first put forward by Mariotte, but it was far from correct. Later Abbé Bossut, who along with Condorcet and d'Alembert conducted a great many resistance experiments, remarked on the unsatisfactory state of affairs as:

'It is very difficult to determine in an exact and practical manner the laws of the impact of fluids. A satisfactory theory on this subject has not yet been found. In the one [theory] ordinarily used, and which has the advantage of being rather simple, it is supposed that the fluid is composed at each instant, of an infinity of parallel jets in the direction of its motion which impinge on the object without interfering with each other. This actually cannot take place and leads in certain cases to results too remote from actuality to be permissible. Nevertheless I have undertaken to expose here this theory despite its imperfections, for two reasons: one is to facilitate to my readers the understanding of several works on naval architecture, of which it is the basis, and the other is that it can be employed, without fear of too great errors, as I have ascertained myself by experiments, in the calculation of water-wheels. . .'[11]

The results of De Pitot's experiments were summed up as follows:

'Now, according to experiments, the perpendicular percussion of an infinite fluid against a plane at rest is essentially equal to the weight of a column of this fluid, which has for base the surface impinged upon and for height the height of the velocity with which this impact takes place. Thus if P is this percussion, s^2 the surface, h the height due to the velocity, and w the specific weight of the fluid, we have approximately

$$P = ws^2h,$$

h can be determined from the laws of fall of heavy bodies.'[11]

Even allowing for the fact that De Pitot's understanding of the flow process was imperfect, the new instrument was a splendid invention. Through its use it became possible to determine velocities in a river at various depths, and prove the fallacy of the parabolic velocity distribution (less velocity at the surface with the maximum velocity at the bottom). So far as the development of hydrology is concerned this, by itself, is a great milestone.

DANIEL BERNOULLI AND THE ENERGY EQUATION

The Swiss family of Bernoulli were prolific mathematicians. Johann Bernoulli (1667–1748) succeeded his elder brother Jakob as the professor of mathematics at Basle, and prior to the appointment, he was a professor at Groningen in the Netherlands. Johann's son Daniel (1700–1782) taught mathematics at St. Petersburg (presently Leningrad) from 1725 to 1732 and then returned to Basle to teach anatomy, physics, and botany.

So far as hydrology is concerned, the main interest lies in Daniel Bernoulli and his analysis of the pressure–velocity relationships. He started writing his book *Hydrodynamica, sive de viribus et motibus fluidorum commentarii* in St. Petersburg during the early thirties. It was published in Strasbourg, in 1738. It made him a leading mathematician, but it probably made his father jealous. The book *Hydraulics now discovered and proved from purely mechanical foundations* was written by his father in 1743, but under an indicated date of 1732[12] – an obvious attempt to steal Daniel's honours. Johann's book, nevertheless, was highly regarded, and Leonhard Euler (1707–1783), to whom he had sent parts of the manuscript prior to its publication, praised it in glowing terms.

Daniel Bernoulli stated the principle behind his analysis as follows:

'But we must give an account of the principles which we have so often mentioned. Of first importance is 'the conservation of live forces', that is, in my notion, the 'equality between the actual descent and the potential ascent': I will use this last notation because, though it has the same significance as the first, it is found that it is perhaps less shocking for certain philosophers, who get excited at the mere name of 'live forces'.'[13]

Like his predecessors, Huygens and Leibniz, he assumed that the sum of potential and kinetic energies of a freely falling body is constant (the terms potential and kinetic energies had not yet been coined).

The main goal of Bernoulli was to establish the relationship between pressure and velocity. He said of his theory that it was new as 'it considers at the same time the pressure and the motion (velocity) of fluids'. It is difficult to say who first tried to solve the pressure–velocity relationship, but certainly Bernoulli was one of the first, even though he was unable to derive a general expression. Instead, he solved some special cases.

The 'Bernoulli equation', as it is internationally known today, was not his work. His analysis included, as previously indicated, only the conservation of summation of potential and kinetic energies, and the effect of pressure was not embodied. The Bernoulli equation,

$$\frac{v^2}{2g} + \frac{p}{\gamma} + z = \text{constant},$$

is basically the principle of conservation of energy, and is a very useful integral form of Euler's equation of motion. The question at once arises as to whether it is fair for Bernoulli to be credited with a principle that he did not propound. The answer is: it may not be wholly fair, but since he was one of the pioneers who directed a much-needed attention to the pressure–velocity relationship, the honour bestowed on him by later investigators is largely justified.

CALCULATION OF DISCHARGE BY CHÉZY AND DU BUAT

Antoine Chézy (1718–1798; figure 3), born at Châlons-sur-Marne, studied and later taught for some time at the local parochial school.

Figure 3. Antoine Chézy (by courtesy of École des Ponts et Chaussées).

He entered what was later named Ecole des Ponts et Chaussées, in 1748, and graduated with honours. Later he joined the Ecole (whose first director was Perronet) and retired, a very poor man, in 1790. Through the efforts of Baron Riche de Prony, one of his former pupils, he was appointed the director of the Ecole in 1797, only a year before his death.

The water supply of Paris was not in a very efficient condition during the latter half of the eighteenth century. To alleviate that circumstance it was recommended by a committee that additional water should be brought from the river Yvette. The city administrators, in 1768, entrusted Perronet and Chézy with the design aspects of the project. Chézy was to determine the cross-section of the channel and its corresponding discharge. In the absence of any acceptable methodology, he had to carry out his own investigations. His final recommendations were handed over to Perronet. The Chézy formula, according to Prony, was established in 1775. The original report on the Canal de l'Yvette, however, was unavailable to him, and it was not until 1897, that the American civil engineer Clemens Herschel, found it among the files of the Ponts et Chaussées. He later had it published.[14]

The following is a translation of the relevant passages that established what is now the famous Chézy formula:

'When we have a flow of water to convey, either to procure some at a place where there is none, or to drain a territory which has too much of it, we should always cause a maximum quantity to flow with the least possible slope.

After having designed a ditch or channel, and having adjusted and regulated its slope, it is very interesting to know if the section of this channel will be sufficient to conduct the water which is to flow in it. To know this, it is necessary to know the speed which the water will have in the ditch, which we will suppose to have a uniform slope.

This is not now a question of initial or momentary velocity, which may be very great if it is caused by a head of water, or very little at first, if it is caused by no other force than that of gravity and the slope of the ditch. Whatever be the initial velocity, it will diminish or augment quite rapidly, and will become that uniform and constant velocity which is due to the slope of the ditch and to gravity, whose effect is impaired by the resistance of friction against the sides of the ditch. It is this velocity which we are now to learn, at least approximately. The question thus proposed presents its own solution, for it is evident that the velocity due to gravity alone, which acts continuously (abstracted from the velocity which may have come from any other cause, and which, being dissipated, no longer concerns the question), this velocity due to gravity is only uniform when it is no longer accelerated, and gravity does not cease to be accelerated, except when its action upon the water is equal to the resistance occasioned by the wetted perimeter of the ditch, but its resistance is as the square of the velocity, on account of the number and force of the particles moving in a given time; it is also as the length of the wetted perimeter of the ditch. The resistance of the air against the surface of the water may be neglected.

If we call V velocity, and the wetted perimeter P, the resistance of friction will therefore be as VVP.

On the other hand, the effect of gravity is as the area of the section of the flowing water and as the slope of the ditch, or as the height which it falls for each toise (6.4 ft) of length.

Calling now the area of the section a, and the slope of the ditch h, the effect of the gravity will be as ah.

This granted, if by a good observation one knew the slope of a ditch . . $\quad H$
the area of the section of the flowing water . . . . $\quad A$
its velocity . . . . $\quad V$
and part of the perimeter of the section of the flowing water touching the
confines of the ditch . . . . $\quad P$
it would be easy to find the velocity . . . . $\quad v$
of the flowing water of another ditch of which one would know the slope . $\quad h$
the area of the section . . . . $\quad a$
and the quoted portion of perimeter . . . . $\quad p$
for one would have the proportion

$$VVP : AH : : vvp : ah$$

whence, ... $VVP \cdot ah = vvp \cdot AH$

$$v = V \sqrt{\frac{a h P}{AHp}} \cdot \text{'14}$$

To prove the formula thus derived, Chézy conducted two experiments in the Courpalet canal in the forest of Orleans and in the river Seine, during the months of September and October of 1769. Both the stretches chosen were 'as straight and as uniform as possible', and measurements were taken on a calm day. Velocities were determined by the float method, using a ball of wax, and the surface velocity was assumed to be the mean velocity. Details of his investigation are shown in the following table.

TABLE

Details of Chézy experiments.

	Courpalet canal	River Seine
Date of experiment	September 23, 1769	October 7, 1769
Velocity	0.468 f.p.s.	2.576 f.p.s.
Slope	0.07224 per 1000	0.1157 per 1000
Area	7.265 sq. ft	3066 sq. ft
Wetted perimeter	7.679 ft	338.988 ft

If the proposed Canal de l'Yvette had a trapezoidal cross-section (5 ft width at bottom and 6 ft at top), slope of 0.2083 ft per 1000 ft, and 5 ft depth of flow at full-supply discharge, the velocities, as calculated by his formula, would be as follows:

from 1st experiment – 1.14 f.p.s. $(C = 56.5)$,

2nd experiment – 1.599 f.p.s. $(C = 79.3)$.

Chézy decided that the first result was too little and the second was too large because of inequalities in the river slope as well as the cross-sectional areas. He reasoned that the velocity in the proposed canal would be slightly more than 1 f.p.s. with a discharge of 33.3 cu. sec, an amount which is more than sufficient if the desired flow is around 17 to 24 cu. sec.

If from the two experiments, Chézy had decided to calculate the velocity of the Seine from that of the Courpalet canal, he would have found it to be 1.83 f.p.s. instead of 2.576 f.p.s. The reason of the discrepancy was that the equation was derived by simple comparison of flow conditions in the two streams, and hence, to obtain good results, they should have very similar characteristics.

In a later memorandum (dated 1776), he simplified[15] his equation to facilitate quick calculations as follows:

$$v = 272 \sqrt{\frac{ah}{p}} \qquad \text{in French units}$$

or

$$v = 57.3 \sqrt{rs} \qquad \text{in English units.}$$

However, he realized that the numerical factor was not constant in every case – in fact, his own calculations showed it to vary from one river to another.

Unfortunately Perronet did not include Chézy's analysis in his report on the Canal de l'Yvette even though he used his results. Thus, the extraordinary piece of work gathered dust in the archives until it was mentioned by Girard in one of his memoirs, in 1803, and by Prony a year later in another memoir. Strangely enough, it attracted more attention in Germany than in France, and it was eventually published by Herschel, in 1897, in the United States! Pierre Louis Georges Du Buat (1738–1809), a contemporary of Chézy, was born in Tortizambert in Normandy. He became a count on the death of his older brother. The title, however, did him no good as it was only two years before the French revolution, and he had to leave his native land in 1793. His property was confiscated, and when he returned in 1802, he could recover only a small portion of it.

Du Buat was educated in Paris, and worked as a military engineer during the period of 1761 to 1791. Over a period of years, he carried out a number of experiments under the sponsorship of the French Government. The first edition of his work, entitled *Principes d'hydraulique vérifié par un grand nombre d'expériences, faites par ordre de Gouvernment*, was published in 1779. The book was enlarged into two volumes in 1786, and was subsequently published posthumously (in 1816) in three volumes. His work was considered so valuable that it was translated into German twice, as well as into English, which version is said to have been praised by no less a person than George Washington.[16]

Du Buat was evidently not happy with the existing state of the science of hydrometry as he said in the preface of the second edition of his work:

'We are still, after so many centuries, in almost absolute ignorance of the true laws to which the motion of water is subjected: hardly after 150 years of experiment have we discovered the quantity and the velocity of the flow of water from any orifice whatever. All which relates to the uniform motion of the streams which water the surface of the earth is unknown to us, and to have any idea of what we know it is only necessary to glance at what we are ignorant of.

To estimate the velocity of a river of which one knows the width, the depth, and

slope; to determine to what height it will rise if it receives another river in its bed; to predict how much it will fail if one diverts water from it; to establish the proper slope of an aqueduct to maintain a given velocity, or the proper capacity of the bed to deliver to a city at a given slope the quantity of water which will satisfy its needs; to lay out the contours of a river in such a manner that it will not work to change the bed in which one had confined it; to calculate the yield of a pipe of which the length, the diameter, and the head are given; to determine how much a bridge, a dam, or a gate will raise the level of a river; to indicate to what distance backwater will be appreciable, and to foretell whether the country will be subject to inundation; to calculate the length and the dimensions of a canal intended to drain marshes long lost to agriculture; to assign the most effective form to the entrances of canals, and to the confluences or mouths of rivers; to determine the most advantageous shape to give to boats or ships to cut the water with the least effort; to calculate in particular the force necessary to move a body which floats on the water. All these questions, and infinitely many others of the same sort, are still unsolvable: who would believe it? . . . Everybody reasons about Hydraulics, but there are few people who understand it . . . For lack of principles, one adopts projects of which the cost is only too real but of which the success is ephemeral; one carries out projects for which the goal is not attained; one charges the state, the provinces, and the communities with considerable costs, without fruit, often with loss; or at least there is no proportion between the cost and the advantages which result therefrom.
The cause of such a great evil, I repeat, is the uncertainty of the principles, the falsity of theory which is contradicted by experience, the paucity of observations made up till now, and the difficulty of making them well.'[17]

Du Buat reasoned that in case of uniform flow, the accelerative force causing the motion should be equal to the sum of the resistances encountered due to viscosity as well as boundary friction.[18,19] The same principle, as already has been noted, was advanced by Guglielmini, but the Frenchman was the first to express it analytically. The principles on which the formulae of Chézy or Du Buat were based, were very nearly identical. Du Buat reasoned that the resistance is proportional to the square of the velocity or equal to $\dfrac{V^2}{m}$, where m is the constant of proportionality; which should be equal to the gravitational force in the direction of flow, i.e.

$$\frac{V^2}{m} = gS, \text{ or } V^2 = mgS.$$

This, obviously, is a form of the Chézy equation. He realized that the term m will be constant only for the same section of the canal and it should vary, in different sections, with the first power of the hydraulic radius R or A/P. He tried to fit an equation of the type $V = \sqrt{gRS}$ to his numerous experimental data, and suggested the following cumbersome equation:

$$V = \frac{\sqrt{243.7\,g}\ (\sqrt{R} - 0.1)}{\sqrt{1/S} - \log\sqrt{1/S} + 1.6} - 0.3\,(\sqrt{R} - 0.1)\ \text{pouces/sec}$$

$$= \frac{297\,(\sqrt{R} - 0.1)}{\sqrt{1/S} - \log\sqrt{1/S} + 1.6} - 0.3\,(\sqrt{R} - 0.1).$$

It is to be noted that the equation does not take into account the surface resistance at all; not because Du Buat neglected it, but because of his abstract conception that a thin layer of water adhered to the boundaries, and the only effective friction taking place was between the fluid molecules themselves. He said:

'Considering how the water itself prepares the surface over which it flows, one can see that the different boundary materials will not have an appreciable influence on the resistance. We have not, as a matter of fact, found any variation in the friction which one could attribute to this cause, in the different cases when water flowed over glass, lead, iron, wood or various kinds of earth.'[17]

If this was the beginning of the present day boundary layer theory, Du Buat was not aware of that fact!

The impact of his equation on the eighteenth and the nineteenth century hydrologists was enormous. It gave excellent results so long as it was used within the range of his experiments, and according to Dugas, it was the best algebraic expression available for the next seventy years – within its limits.[20] However, it failed when the conditions were beyond the range of his experiments, and thus, could not be the all-purpose equation its originator wanted it to be. Besides, the expression as it stood was too cumbersome to use, and even though it was arrived at independently, its introduction came four years after Chézy's pioneer work.

PAOLO FRISI

Paolo Frisi (1727–1784) was born in Milan, and his early studies were directed by the Church of St. Barnabas in his native city. He became a professor of mathematics at the University of Milan, and was a member of the most of the major scientific societies of his time. His patrons included Maria Theresa, Catherina II, and Joseph II. His reputation in Italy, in the field of hydrometry, was so high that plans for all major works executed during his time were submitted to him for his comments. His book *Del modo di regolare i fiumi e i torrenti*[21,22] was published in 1762, and was well received. It was considered to be of such importance that the British government paid for all the expenses incurred for its English translation so that it could be made available to the British engineers engaged in irrigation and river regulation works in India.

The book admirably discusses the previous works in the fields of hydrometry and open channel flow by the Italian school – Castelli, Viviani, Zendrini, Manfredi, Poleni, Grandi, and especially Guglielmini.

Frisi was confident that his opinion on the origin of rivers and springs was correct:

'In short, all the phenomena of floods; the laws, by which they increase and diminish; the substance, which they sweep along with them; all clearly point out, that they derive their origin from the rains that fall on the declivities of the mountains and into the beds of the rivers. Seeing then, the greatest quantity of water, as has been before observed, is brought down by the rivers in the time of their high and moderate floods, it would be unreasonable not to admit that the waters, when low, have the same origin.'[22]

Frisi was wrong about the velocity distribution pattern in an open channel. However, he was in good company so far as the parabolic distribution law was concerned, with people like Zendrini (1679–1747), Lecchi (1702–1776), Michelotti (1710–1777), and Lorgna (1730–1796). He discussed at length the relevant passages from Zendrini's[23] and Father Grandi's[24] works, and erroneously concluded that:

'It must appear to be sufficiently ascertained that the velocities of water, though arising from different causes, either from the free fall, or from the pressure of the higher waters, have only one law, and are proportional to the square roots of the heights, either actual or effective; that is, they are in proportion to the square roots of the actual and absolute heights of the sections, when the surface of the water has no perceptible motion; and, when the motion of the surface is perceptible, they are proportional to the square roots of the actual heights augmented by the height due to the velocity of the surface.'[25]

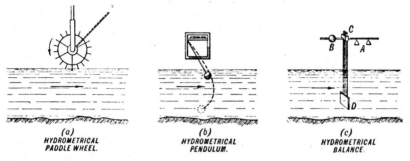

(a)
HYDROMETRICAL
PADDLE WHEEL.

(b)
HYDROMETRICAL
PENDULUM.

(c)
HYDROMETRICAL
BALANCE.

Figure 4. Velocity measurement devices (after Leliavsky).

Leliavsky[26,27] has suggested that the origin of the parabolic velocity distribution lies in the methods used for measuring velocities. The 'hydrometric pendulum' (figure 4) evaluated velocities from the angle of inclination of the pendulum, i.e., greater the angle of inclination, higher is the velocity. Such a device, according to Leliavsky, will overestimate velocities at lower levels as the curvature of the cord is neglected. He further stated that the parabolic velocity distribution theory was so firmly established by that time that any reading which did not conform to the theory was rejected as an error of measurement, an unfortunate human tendency. While Leliavsky is probably correct in those opinions, his drawing thereof (figure 4b) appears to exaggerate the circumstances. When a ball is lowered to near the bottom, the suspension cord could not assume such an upstream curve as he has shown unless the deeper currents are actually flowing in the upstream direction (a condition which occurs in some instances on tide-affected streams when there is an incoming tide). But in a stream where there is no such reverse current, and where the ball is not resting on the bottom, neither the ball nor any part of the lower cord could come to rest at a point that is upstream from any other portion of the cord.

Frisi also discussed the use of floats and paddle wheels (figure 4a) for measuring surface velocities. The number of revolutions of the wheel, whose vanes touched the surface of the stream, per unit time was taken to be an indication of the velocity.[28] Leliavsky[26,27] also suggested that use of such a device would result in underestimation of velocities because of the effect of water lifted by the vanes. Probably Frisi's greatest contribution to hydrologic knowledge was his contention about the limitations of purely theoretical approach to open channel flow. He pointed out:

'One single reflection is sufficient to show that all hydraulic problems are beyond the reach of geometry, and of calculus. The difficulty of all problems is increased in proportion to the number of the conditions, of the cases, and of the differences which are stated. Thus, mechanical problems become so much the more complicated as the number of bodies, whose motions are sought, and which act in any way on each other, is augmented . . . Then, in a fluid mass, which moves in a tube, or in a canal, the number of bodies acting together is infinite; whence it follows, that to determine the motion of each body is a problem depending upon an infinity of equations, and which it is of course beyond all the powers of algebra to reach.'[29]

Thus, concluded Frisi, hydrometry is a branch of physics rather than of mathematics,[30] and that idea was later echoed and re-echoed from various parts of the world.

Paolo Frisi did not contribute any significant original concepts to the development of hydrology, but his work was an excellent compilation on the subject, and it did much to disseminate the available knowledge all over the world, which, in itself, is a praiseworthy achievement.

During the last quarter of the eighteenth century, the prestige of the Italian school gradually dwindled. For some time, previous to the above-mentioned period, no significant new ideas were put forward, even though a series of treatises on rivers and canals – all of them discussing the existing state of hydrologic knowledge – were printed.

GIOVANNI BATTISTA VENTURI

Giovanni Battista Venturi (1746–1822) was a professor of Natural Philosophy at the University of Modena and later at Pavia.[31] He was a noted Italian civil engineer, builder of bridges,[32] and obviously a highly talented experimental hydraulician. His experiments were carried out in the physical laboratory of the University of Modena over a long period of time, and the results of his investigations were later collected and published at Paris, in 1797, as *Recherches expérimentales sur le principe de la communication latérale du mouvement dans les fluides*.[33]

Venturi experimented with various types of constrictions and expansions in open channels as well as with orifices and short tubes of various shapes.[34] Clemens Herschel was primarily instrumental in making Venturi well-known by associating his name with a particular type of boundary-profile now universally referred to as the Venturi flume. But, it must be realized that Venturi was by no means the first man to experiment with conduits of that particular shape – many, notably Bernoulli and Borda, had been interested in them before him.

Venturi's book is a straightforward report of his investigations, and one immediately becomes aware of its lack of analyses or computations. He discussed the effect of eddies in open channels and pointed out similar phenomena can be seen in the motion of the atmosphere. He showed that formation of eddies is a cause of retardation of the current and hence of the discharge. One of the interesting ideas put forward by Venturi was the use of the principle of the hydraulic

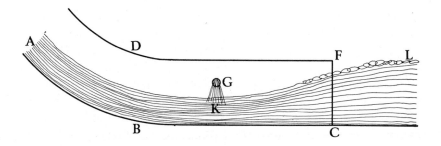

Figure 5..Venturi's use of the principle of hydraulic jump for drainage problems.

jump for drainage problems as shown in figure 5. His explanation was:

'water at F tends to return and descend along FK; but the current, by its lateral action, constantly carries it away, and does not permit it to slide down to K. If an opening G be made in the lateral sides of the tank, the waters from lands lower than the current of the inferior stream FL may be drained off.'[35]

The principle was used by Venturi in an actual case, and, according to him, it was very successful.

MEASUREMENTS OF EVAPORATION

After Halley, the first noteworthy measurements of evaporation were taken by Dobson,[36,37] who maintained a record of the rainfall, evaporation, and temperature for four consecutive years, from 1772 to 1775. Dobson used a site overlooking Liverpool, 75 ft above sea level, on rising ground having free exposure to sun, wind, and rain, with the measurements being taken at the middle of a grass-covered plot. He used:

'two well-varnished tin vessels; one of which was to serve the purpose of rain-gauge; the other to be employed as my evaporating vessel. The evaporating vessel was cylindrical, twelve inches in diameter and six inches deep. The rain-gauge consisted of a funnel twelve inches likewise in diameter, the lower end of which was received into the mouth of a large stone-bottle; and to prevent any evaporation from the bottle, the pipe of the funnel was stopped with grooved cork.'[36]

Water level in the evaporation measurement vessel was kept at 2 in. below the rim. Depending on rainfall or evaporation, water was either taken out or added in order that the level in the vessel would remain constant. Knowing the rainfall and the amount of water added or removed, he could calculate the monthly evaporation. The temperature was measured every day at 2 p.m. by a thermometer attached to a shaded wall. The main defects of such a method were: the difficulty of restoring the water level in the tank to its exact original height, the precise measurement of water added or removed (including that lost by splashing), and the possibility of water being lost by overflowing.

Rodda[38] has estimated the evaporation of the same period from Dobson's data using the equation:

$$E = 0.17T - 7.18$$

where, E = monthly pan-evaporation in in.,

T = monthly mean maximum temperature in °F.

Even allowing for the fact that no factors were included in the equation to compensate for the climatic differences between Wallingford (Rodda) and Liverpool (Dobson), or to account for other controls of evaporation, the values obtained by Dobson seem relatively high. This supposition is substantiated in part by the fact that Pennman[39] estimated the annual average evaporation at Southport to be 26 in.

John Dalton later used Dobson's method to determine evaporation at Kendal for eighty-two days in March, April, May, and June, and found it to be 5.414 in. During the period, the maximum evaporation recorded in a day was little above 0.2 in. Dalton stated that a certain Dr. Hale, from a few experiments conducted, concluded that 6.66 in. of water evaporated annually from 'green ground and moist earth', which according to him 'must be far below truth'. The Bishop of Llandaff had found that:

'in a dry season there evaporated from a grass plot that had been mowed close about 1600 gallons in an acre per day which amounts to 0.07 of an inch in depth; and that after rain the evaporation was considerably more.'[40]

With the help of his friend Thomas Hoyle, Jr., Dalton determined the evaporation at an unspecified site near Manchester from the autumn of 1795. A cylindrical vessel of tinned iron, 10 in. in diameter and 3 ft deep, was used. Two pipes were connected to the vessel – one at the bottom and the other an inch from the top. The vessel was filled for a few in. with gravel and sand and the rest with good fresh soil.

'It was then put into a hole in the ground and the space around filled up with earth, except on one side, for the convenience of putting bottles to the two pipes; then water was added to sodden the earth, and as much of it as would was suffered to run through without notice, by which the earth might be considered as saturated with water.'[40]

Initially, the soil was kept about the level of the upper pipe for some weeks, but later it was below the pipe so as to preclude any water from flowing down the pipe. Moreover, soil at the top was bare during the first year but was covered with grass for the following two years. A regular record was kept of the quantity of rain water which ran off from the surface of the earth through the upper pipe and also the quantity that percolated through the sample to the bottom pipe. Rainfall during the corresponding time was measured by a cylindrical vessel having the same dimensions as the one used for evaporation measurements. Dalton assumed that:

evaporation — rainfall = quantity of water in the two bottles.

From the experiment conducted Dalton concluded that:
(1) the annual evaporation under the circumstances stated was 25 in., (2) quantity of evaporation increases with the rain but not proportionally, and (3) there is no difference between evaporation from bare earth and vegetating grass.

In a subsequent paper[41] Dalton gave the results of his observations of evaporation from the water surface of a cylindrical vessel 10 in. in diameter during the period 1799–1801.[42] In 1802, he also put forward a generalized theory of vapour pressure[43] which provided an excellent basis to estimate the rate of evaporation from water surfaces. The theory is based on the observation that under given conditions evaporation is proportional to the deficit in vapour pressure. Expressed mathematically, it takes the following form:

$$E = C(e_w - e_a),$$

where

E = rate of evaporation in in. per day,
C = a coefficient (depending on various unaccounted for factors affecting evaporation),
e_w = maximum vapour pressure (in mercury),
e_a = actual vapour pressure (in mercury).

The method is still extensively used today except for slight modifications which take into account effects due to wind and/or temperature.

The beginning of experimental research in the field of suppression of evaporation by a film of oil was initiated by Benjamin Franklin (1706–1790). In 1765, he conducted experiments on the spreading of oil on water surfaces in a large pond at Clapham Common in England. In a letter to one William Brownrigg he pointed out that if a drop of oil was placed on a horizontal mirror or a highly polished table, the drop remained in its place, whereas:

'when put on water, it spreads instantly, becoming so thin as to produce prismatic colours, for a considerable space, and beyond them so much thinner as to be invisible, except in its effect of smoothing the waves at a much greater distance.'[44, 45]

Franklin's main interest seemed to have been the use of oil as a method for wave damping, and from the experiments he concluded that the minimum thickness of the film should be about 25 Å.

OTHER EQUIPMENT AND PROCEDURE DEVELOPMENTS

William Heberden was probably the first to study the variation of rainfall with elevation.[46] His interest was first aroused when he studied the results of observations from two identical rain gauges in London which were placed about a mile apart. He found that the rainfall at one station continually exceeded that of the other – not only every month but also almost every time it rained. He reasoned that 'this unexpected variation' must have something to do with elevation because one of them was fixed above all the neighbouring houses whereas the other one was at a considerably lower level. In order to verify his theory, Heberden used two rain gauges – one on the chimney of a house (probably in London) and the other in the garden of the same house. The results confirmed his theory, and he decided to use a third one on the roof of the Westminster Abbey. Observations were taken at monthly intervals for a year, but he was unable to determine the reason for the variation of rainfall. He surmised, erroneously, that:

'Some hitherto unknown property of electricity is concerned in this phaenomenon. This power has undoubtedly a great share in the descent of rain, which hardly ever happens, if the air and electrical apparatus be sufficiently dry, without manifest signs of electricity in the air.'[46]

Reinhard Woltman (1757–1837), a German engineer, spent nearly all his working life in the Department of Ports and Navigable

Waterways in Hannover. In his book *Theorie und Gebrauch des hydrometrischen Flugels*,[47] published in 1790, he described a spoke-

Figure 6. The current meter of Woltman (reconstructed by Arthur H. Frazier).

vane type current meter (figure 6) with a revolution counter to determine river discharge. For many years after Woltman's death, every improved version of his current meter was called a 'Woltman current meter' out of courtesy to him. Hydrology and hydraulic textbooks and many articles have published drawings of those improved models, and have erroneously described them as having been designed by Woltman.

Extensive experimental works were carried out by Michelotti in

Turin and Abbé Bossut (1730–1814) in Paris. Michelotti carried out a series of experiments under the patronage of the King of Sardinia, and his results were published in 1774.[48] Bossut's tests were carried out at the expenses of the French government, and the results were published between 1771 to 1778. Both of them made several experiments on river and canal flow problems.

The concept of the hydrologic cylce was extended by Jean-Claude De La Méthèrie (1743–1817). He explained[49] that the total rainfall is disposed of in three ways: by the first if flows off directly to rivers and canals; by the second it is released through evaporation, transpiration from plants, or moistening of the soil; and by the third it infiltrates to greater depths and provides a source of water for springs.

Probably one of the major contributors from England to water science of the eighteenth century was the Leeds-born John Smeaton (1724–1792), who was responsible for the design of various harbours and drainage works. His main contributions, however, were the tests he made on scale models of water wheels[50] and windmills. Many treatises and papers were written in the field of open channel flow during the eighteenth century, the more important of which have been discussed in this chapter. Other contributors during this period included Pierre Varignon (1654–1722), Bernard Forest De Belidor (1693–1761), Alexi Claude Clairaut (1713–1765), Jean Le Rond d'Alembert (1717–1783), Jean Charles Borda (1733–1749), and Jean Antoine Fabre (1749–1834) in France; Antonio Lecchi (1702–1776), and Antonio Mario Lorgna (1735–1796) in Italy; Christiaan Brünings (1736–1805) in Holland; and John Theophilus Desaguliers (1683–1744) and James Jurin (1684–1750) in England.

CONCLUSION

The main development concerning hydrology in the eighteenth century was in the field of surface water. Considerable experimental results were obtained particularly by the Italian and French hydrologists. The influence of the Italian school was predominant at the beginning of the century but it gradually dwindled towards the latter half, and its place was taken over by the French school.

The reason seems obvious: the Italian hydrologists were writing and rewriting what had already been said before, and, hence, most of such treatises did not introduce any important fundamental or new concepts.

The new 'machine' of De Pitot, although developed on the basis of two erroneous principles, revolutionized the method of measuring velocities in rivers. By its use, the concept of the parabolic distribution of velocity in rivers and canals was proven by actual experiments to be fallacious. This absurd theory had done considerable harm to the natural development of concepts of open channel flow. Undoubtedly, the most influential man of this period was Du Buat, and his influence extended well into the nineteenth century. Along with Borda, he was one of the first scientists to write the efflux equation correctly as $v = \sqrt{2gh}$. But, Du Buat's major claim to fame was his experimental work in the field of quantifying discharge. Using the mass of data collected, he produced an algebraic expression for computing the discharge of open channels. The expression obtained was no doubt cumbersome to use, and is only valid within the narrow range of his experiments, but, nevertheless, it was a very significant development of his time.

Modern students of hydrology are more familiar with the Chézy equation than with that of Du Buat. Both of them were based on almost identical assumptions, but Chézy was the first to produce such a formula. Unfortunately, Du Buat was not aware of Chézy's contribution, nor as a matter of fact, was anyone else till it was published in the nineteenth century. Chézy's equation remained in its most primitive form, whereas Du Buat tried to develop a generalized expression which could be applied universally. Furthermore, Chézy's expression relied on experience for adapting it to specific problems, but the Du Buat formula did not. Looking back, it seems extremely unfair that all students of hydrology or open channel hydraulics should become aware of Chézy's equation, while very few of them are even aware that Du Buat's existed!

REFERENCES

I. CARDWELL, D. S. L., The organisation of science in England. London, William Heinemann Ltd. (1957) p. 10.

2. VALLISNIERI, A., Lezione accademica intorno all'origine delle fontane. Venezia (1715).

3. Quoted by ADAMS, F. D., The birth and the development of the geological sciences. New York, Dover Publications Inc. (1954) p. 453.

4. POLENI, G., De motu aqua mixto libri due. Patavii (1717).

5. DERHAM, W., Physico-theology. London, W. Innys (1713).

6. SWITZER, S., An introduction to a general system of hydrostaticks and hydraulics, vol. 1 (1729) pp. 27–29.

7. PLUCHE, N. A., Le spectacle de la nature. Paris (1732). Translated by J. Kelly, D. Bellamy and J. Sparrow, vol. 3, 3rd ed. London, J. Hodges (1744) pp. 80–81.

8. DE PITOT, H., Description d'une machine pour mesurer la vitesse des eaux courantes et le sillage des vaisseuax. Mémoires de l'Académie Royale des Sciences (1732) pp. 363–376.

9. Ref. 8 was reprinted in La Houille Blanche, no. 8 (1966) 922–936.

10. KIRBY, R. S., Henry De Pitot, pioneer in practical hydraulics. Civil Engineering, ASCE 9 (1939) 738–740.

11. BOSSUT, C., Traité théorique et expérimental d'hydrodynamique. Paris (1786–87).

12. ROUSE, H. and S. INCE, History of hydraulics. Iowa City, Iowa Institute of Hydraulic Research (1957) pp. 91–92.

13. BERNOULLI, D., Hydrodynamica. Argentorati, J. R. Dulseckeri (1738).

14. HERSCHEL, C., On the origin of the Chézy formula. Journal of the Association of Engineering Societies 18 (1897) 363–369.

15. MOURET, G., Antoine Chézy, histoire d'une formule d'hydraulique. Annales des Ponts et Chaussées, II (1921).

16. The statement comes from LAROUSE GRAND DICTIONNAIRE UNIVERSAL, XIX siècle, tome 6, 'D', pp. 1319–1320. My extensive searches have failed to locate an English translation of Du Buat's book which George Washington could have seen.

17. DU BUAT, P. L. G., Principes d'hydraulique, 2 vols., new ed. Paris (1786).

18. NEMÉNYI, P. F., The main concepts and ideas of fluid dynamics in their historical development. The Archives for History of Exact Sciences 2, (1962) 52–86.

19. DU BUAT, P. L. G., Chevalier Du Buat's principles of hydraulics, translated by T. F. De Havilland, 2 vols. Madras, Asylum Press (1822).

20. DUGAS, R., Histoire de la mécanique. Paris, Dunod (1950).

21. FRISI, P., Del modo di regolare i fiumi e i torrenti. Lucca, G. Riccomini (1762).

22. FRISI, P., A treatise on rivers and torrents, translation of ref. 21 by J. Garstin. London, Longman, Hurst, Rees, Orme and Brown (1818) p. 5.

23. ZENDRINI, B. and E. MANFREDI, Leggi e fenomeni, regolazioni ed usi delle acque correnti. Venezia (1741).

24. GRANDI, G., Del movimento dell'acque trattato geometrico. In: Raccolta d'autori chi trattano del moto dell'acque. Firenze (1723).

25. FRISI, P., *op. cit.*, pp. 56–57.

26. LELIAVSKY, S., Historic development of the theory of the flow of water in canals and rivers. Engineer, no. 1 (1951) 466–468.

27. LELIAVSKY, S., River and canal hydraulics. London, Chapman and Hall (1965) pp. 3–5.

28. *Ibid.*, p. 54.

29. *Ibid.*, pp. 47–48.

30. FREEMAN, J. R., ed., Hydraulic laboratory practice. New York, American Society of Mechanical Engineers (1929) p. 13.

31. BARDSLEY, C. E., Historical resume of the development of the science of hydraulics, Publication no. 9, Oklahoma A & M College, Stillwater (1939) p. 13.

32. RICHARDSON, C. G., The 50th year of a great invention. Water Works Engineering *90* (1937) 1326.

33. VENTURI, G. B., Recherches expérimentales sur le principe de la communication latérale du mouvement dans les fluides. Paris (1797).

34. KENT, W. G., An appreciation of two great workers in hydraulics: Giovanni Batista Venturi and Clemens Herschel. London, Blades East Blades (1912).

35. VENTURI, G. B., Experimental inquiries concerning the principle of the lateral communication of motion in fluids, translation of ref. 33 by W. Nicholson, 2nd ed. In: Tracts on hydraulics, edited by T. Tredgold, London, J. Taylor (1826) pp. 163–164.

36. DOBSON, D., Observations on the annual evaporation at Liverpool in Lancashire; and on evaporation considered as a test of the dryness of the atmosphere. Philosophical Transactions of the Royal Society of London *67* (1777) 244–259.

37. BISWAS, ASIT K., Experiments on atmospheric evaporation until the end of the eighteenth century. Technology and Culture *10* (1969) 49–58.

38. RODDA, J. C., Eighteenth century evaporation experiments. Weather *18* (1953) 266.

39. PENNMAN, H. L., Evaporation over the British Isles. Journal of the Royal Meteorological Society *75* (1950) 372–382.

40. DALTON, J., Experiments and observations to determine whether the quantity of rain and dew is equal to the quantity of water carried off by the rivers and raised by evaporation; with an enquiry into the origin of springs. Memoirs, Literary and Philosophical Society of Manchester, vol. 5, part 2 (1802) pp. 346–372.

41. DALTON, J., Meteorological observations. Memoirs, Literary and Philosophical Society of Manchester, vol. 5, part 2 (1802) pp. 666–674.

42. MANLEY, G., Dalton's accomplishment in meteorology. In: John Dalton and the progress of science, edited by D. S. L. Cardwell. Manchester, Manchester University Press (1968) pp. 140–158.

43. DALTON, J., Experimental essays on the constitution of mixed gases; on the force of steam or vapour from water and other liquids in different temperatures, both in a torricellian vacuum and in air, on evaporation, and on

the expansion of gases by heat. Memoirs, Literary and Philosophical Society of Manchester, vol. 5, part 2 (1802) pp. 536–602.

44. CROWTHER, J. G., Famous American men of science, 1st ed. London, Secker and Warburg Ltd. (1937) pp. 101–105.

45. GOODMAN, NATHAN M., The ingenious Dr. Franklin. Philadelphia, University of Pennsylvania Press (1956) pp. 188–197.

46. HERBERDEN, W., Of the different quantities of rain, which appear to fall, at different heights, over the same spot of ground. Philosophical Transactions of the Royal Society of London 59 (1769) 359–362.

47. WOLTMAN, R., Theorie und Gebrauch des hydrometrischen Flugels. Hamburg (1790).

48. MICHELOTTI, P. A., Dissertatio de separatione fluidorum in corpore animali. Ventiis (1721).

49. DE LA METHERIE, J. C., Théorie de la terre, vol. 4, 2nd ed. Paris, Chez Maradan (1797) pp. 457-479.

50. CARDWELL, D. S. L., Power technologies and the advance of science, 1700–1825. Technology and Culture 6 (1965) 188–207.

13

The nineteenth century

INTRODUCTION

The rapid increase of knowledge in the field of hydrology during the nineteenth century is indeed remarkable. The experimental methods, that were successfully pioneered by Perrault, Mariotte and Halley in the seventeenth century had already taken firm roots, and, undoubtedly, the major developments during the nineteenth century were in the fields of ground water hydrology and surface water measurements. At the beginning of the century, the French School was still the leading one in the field of hydrology and hydraulics, but it had a rather bad experience due to the French Revolution. The Revolutionary Government suspended the Académie des Sciences in 1793, executed scientists like Lavoisier, Bailly and Cousin, and drove Condorcet to suicide. Fortunately for science, the authorities quickly realized their mistakes. Men like Du Buat, who had to flee from France to save his life, later returned to their native land, and continued to carry on their admirable work.

VELOCITY-DISCHARGE FORMULAE

The Chézy equation remained almost unknown until 1897 the end of the nineteenth century, the prevailing stream-velocity equation was that which had been conceived by Du Buat. The main problems of this period were concerned with the function relation-

ship between velocity (and hence, discharge) and the frictional resistance. Many different theories were advanced and they produced
numerous formulae that attempted to relate velocity, hydraulic
gradient and the hydraulic radius. Notable among the contributors
were Baron Gaspard Clair François Marie Riche De Prony (1755–
1839) and Pierre Simon Girard (1765–1836) from France, and
Johann Albert Eytelwein (1764–1848) from Germany.

De Prony, born in Chamelet near Lyons, was associated with several
engineering works in France and Italy. He was a student of Chézy,
and later received an appointment as Director of the Ecole des
Ponts et Chaussées. Girard was a native of Caen, and a graduate
of the Ecole des Ponts et Chaussées. He took part in Napoleon's
Egyptian campaign, and was later appointed water commissioner
of Paris. The German, Eytelwein, was born in Frankfurt. He was
connected with the construction of various river regulating works
and harbours. He translated many books on hydraulics into German, including that of Du Buat. His book *Handbuch der Mechanik
fester Körper und der Hydraulik*[1, 2] was published in 1801. His later
work on hydrostatics,[3] published in 1826, had considerable influence
on the hydrologists and hydraulicians of his country. It was accepted as a standard work in that field well into the early part of
the twentieth century.

The basic theory of all the early nineteenth century discharge
formulae was expressed in a paper[4] published in 1800 by Charles
Augustin De Coulomb (1736–1806). De Coulomb fastened discs
of various sizes onto the lower end of a brass wire, and immersed
them into different liquids. The law of resistance was determined
from the rate of damping of the rotational oscillation of the discs.
From his elaborate experimentation, De Coulomb concluded that
the resistance can be represented by a function containing two
terms, one of which varied with the first power of the velocity, and
the other with its second power.

In 1803, Girard[5] was the first to apply De Coulomb's law of resistance to the flow of water in rivers and channels, but he used the
same value of the numerical coefficient for the both powers of the
velocity, i.e.,

$$gRS = C(V + V^2).$$

He made only cursory attempts to test the validity of the equation and of the value of the constant C.

The following year, De Prony discussed a series of formulae[6] for evaluating the flow of water in open channels as well as through orifices. He concluded that De Coulomb's law of resistance was a part of the infinite series:

$$C + aV + bV^2 + cV^3 + \ldots$$

The mean velocity V was determined from the surface velocity U,

$$V = 0.816458U \text{ (approximately } \frac{4}{5}U).$$

De Prony,[7] unlike Girard, used two separate coefficients in his resistance formula:

$$RS = 0.0000444499V + 0.000309314V^2.$$

Another pair of coefficients was suggested in 1818 by Eytelwein:[8]

$$RS = 0.000024V + 0.000366V^2.$$

Those two coefficients were assumed to be independent of the extent of boundary roughness. While De Prony's coefficients were widely accepted in France, Eytelwein's formula received more favour elsewhere. For higher velocities, the latter can be reduced to the Chézy type by assuming the resistance to be proportional to the square of the velocity, i.e.,

$$V = 50.9\sqrt{RS}.$$

The equation, with the coefficient 50, became quite popular in Italy in the 1830's. There it was known as Tadini formula. Courtois[8] expressed his approval of it in 1850.

Also, in 1845, Lahmeyer[9] tried to provide for the effect of curvature in the channels. For a radius r and width w, the equation he suggested was:

$$\frac{RS}{V^{3/2}} = 0.0004021 + 0.002881\sqrt{\frac{w}{r}}.$$

For a straight channel $r = \infty$, and hence,

$$V = 49.87V^{\frac{1}{4}}R^{\frac{1}{2}}S^{\frac{1}{2}}.$$

The following year De Saint-Venant[10] offered another formula with tables for facilitating quick computations:

$$V = 60(RS)^{11/21}.$$

His equation, however, does not seem to have been used very much.

Eytelwein analysed the experimental results reported by Leonardo Ximenes (1716–1786) of Italy and those of Christiaan Brünnings (1736–1805) of Holland to obtain velocity distribution profiles – but the results turned out to be contradictory. He finally suggested a linear variation. The velocity v at a depth d from the water surface was expressed by:

$$v = (1 - 0.0127d)U.$$

Darcy and Bazin

Henry Philibert Gaspard Darcy (1803–1858) was born in Dijon, and was educated in Paris. Several years prior to 1850, Darcy was concerned with the design and the construction of a public water supply system for the municipality of Dijon. The detailed report on the project,[11] published in 1856, contained technical information as well as historical background, legal considerations, and a series of appendices. The venture was so successful that he was later retained by the city of Brussels to advise on a similar project.

Around 1855, a growing nervous disorder made Darcy relinquish everything but his work on hydraulics. He had an able assistant[12] in none other than Henri Emile Bazin (1829–1917). When Bazin had been stationed in Dijon, in 1854, he had requested a transfer to Darcy's staff. Between them, they set up the most elaborate sets of experiments ever to have been conducted in a laboratory up to that time. They were rather fortunate in that the French Second Empire Government sponsored their outstanding work primarily to exhibit their liberal tendencies. Bazin completed the experimental work with remarkable skill in 1860, two years after the death of Darcy. In 1875, he was made an engineer-in-chief, and by 1886 he was promoted to inspector-general. He retired in 1900 when he had become unable to obtain further research funds to carry out his work. He died in 1917.

The book *Recherches hydrauliques*[13] was published in 1865 under the joint authorship of Darcy and Bazin. The velocity distribution profiles for various types of channels as presented therein were obtained by using an improved version of the Pitot tube, made by Darcy. (It was somewhat similar to present models.) Their experiments revealed that the maximum velocity in natural rivers

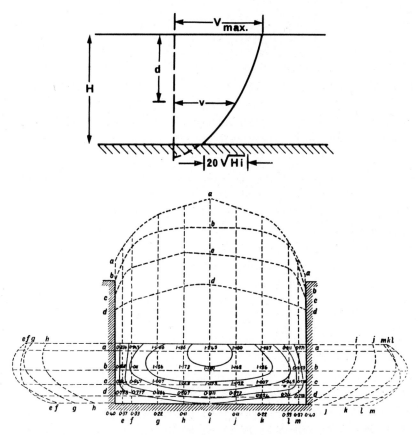

Figure 1. (a) Velocity distribution profile of Bazin. (b) Velocity distribution in narrow channels (from *Recherches hydrauliques* by Bazin).

and in wide channels occurred at the surface. In the centre of wide water courses, where side effects are nil, velocity v at a depth d from the surface (figure 1) was found to be:

$$\frac{V_{max} - v}{H_i} = 20\left(\frac{d}{H}\right)^2,$$

where the linear measurements were expressed in metres. In narrow canals having a width less than five times the depth the maximum velocity was found to occur somewhat below the surface (figure 1). Darcy and Bazin used a 596.5 metre long channel for their experi-

ments. Water for the channel came from the Burgundian canal near Dijon, and it discharged into the river Ouche. The channel had rectangular, trapezoidal, triangular and semicircular sections with various types of linings.
They believed that:

'if there exists an analytical law including all cases it must necessarily be very complicated, and the knowledge of the laws of the motion of fluids is too little advanced to allow us to hope for their discovery at present. In the actual state of the science we must limit ourselves to the search for empirical formulae sufficiently accurate for practical purposes and calculations by which are easy.'[13]

From the experiments conducted, they proposed the following relationships:

$$RS = (a + \frac{b}{R})V^2.$$

This differed from the De Prony equation which was of the type:

$$RS = (a + \frac{b}{V})V^2.$$

The Darcy–Bazin formula attracted considerable attention, but the use of the two variable coefficients limited its practical application. Realizing that difficulty, Bazin proposed[14] in 1897 another equation which contained only one variable coefficient. It is known as the Bazin formula in contradistinction to the previous Darcy–Bazin formula. According to it:

$$C = \frac{157.6}{1 + (1.81\gamma/\sqrt{R})} \quad \text{(in f.p.s. units)}$$

where γ is the rugosity factor which reflects the characteristics of the channel.
Darcy and Bazin also conducted a series of experiments on the flow of water through orifices and conduits. A summary of the development of their various pipe-flow formulae is contained in a series of articles by Davies and White.[15]

Figure 2. Andrew Atkinson Humphreys.

Humphreys and Abbot

Andrew Atkinson Humphreys (1810–1883; figure 2) and Henry Larcom Abbot (1831–1927; figure 3) were both graduates of West Point, and were responsible for the Mississippi delta survey of 1851 to 1860. This was by no means the first river survey, but it was certainly the most extensive that had been undertaken up to that time. Humphreys was initially selected to conduct the survey.

Figure 3. Henry Larcom Abbot.

His intense devotion to the project, and the long hours he devoted to the field work caused him to become seriously ill. While recovering therefrom he was sent to Europe to study developments in water resources field. In 1857, Abbot was appointed as his assistant. Their *Report upon the physics and hydraulics of the Mississippi river*[16] was published in 1861, and was immediately a 'best seller'. The monumental Report had 610 pages, and parts of it were soon translated into many languages. Kolupaila[17] has listed 39 reviews and discussions of it, of which 18 were in foreign languages. It contained a review of the state of the art on river hydraulics which, according to the authors were 'partly original and partly compiled' from similar reviews by Rennie, Lombardini, Storrow and others, and from various encyclopedias.[18]

Humphreys and Abbot used double floats almost exclusively for velocity measurements, and found that for the same stages, the discharge could vary as much as 20%, depending on whether the stage was rising or falling. The mean values were selected for presenting stage–discharge relationships. They proposed a rather complicated formula for determining discharge in a water course. It was based on observations from channels as large as that of the mighty Mississippi to the small artificial channels used by Du Buat. The main importance of the formula was that it was the first of a series of formulae supposed to be comprehensive and accurate. The formula did not contain any roughness term. For that and other reasons, it failed to gain much popular acceptance. Undoubtedly, the main significance of the entire project was its thoroughness.

Ganguillet and Kutter

The two Swiss engineers, Emile Oscar Ganguillet (1818–1894; figure 4) and Wilhelm Rudolph Kutter (1818–1888), were considerably interested in the problems of open channel flow, and carried out a series of experiments in the Swiss mountain streams. Ganguillet was the chief engineer of the Department of Public Works at Berne – Kutter was a member of his staff.

Ganguillet and Kutter were invited by Humphreys and Abbot to check the validity of their (Humphreys' and Abbot's) formula. When they found that it was valid only for streams with gentle

Figure 4. Emile Oscar Ganguillet.

slopes, they tried to develop a formula that would apply to the
discharge in all types of channels. In doing so they used observations

from their own investigations as well as from as many other reliable sources as they could find. On the basis of their analysis, they concluded that:

'The two formulae are equally not entitled to general application. That of Bazin is indeed as inapplicable to the Mississippi as that of Humphreys and Abbot is to channels with steep slopes; but it [Bazin's] contains the basis of a formula which can generally be applied, simply by introducing the effect of slope, while the American formula cannot be thus generalized.'[19]

In f.p.s. units, the expression proposed for the 'C' in the Chézy equation was:

$$\frac{41.65 + 0.00281/S + 1.811/n}{1 + (41.65 + 0.00281/S)\, n/\sqrt{R}}.$$

This equation was an immediate success, and the 1869 journal which contained the article was quickly sold out. In 1877, the paper was enlarged and published as a book[19] which was soon translated into several different languages. According to Manning, the English translation of the article, available in 1876, was 'received generally with great favour'.

The complexity of the equation was justified by its originators on the basis that 'any formula that would possess an adequate claim to universal utility must necessarily be very complicated'. Manning was one of the first men to point out the dimensional non-homogeneity of the expression. Later, Forchheimer commented that it would have been more accurate if the early records of the Mississippi river discharge had been excluded when the averages were computed.

In spite of its inherent disadvantages, Ganguillet and Kutter's formula has received a wide acceptance all over the world. In 1905, Merriman considered that,

'It is to be regarded as a formula of great value, so that no design for a conduit or channel should be completed without employing it in the investigation, even if the final construction be not based upon it.'[20]

Robert Manning

Robert Manning (1816–1897; figure 5), an Irishman, was born

Figure 5. Robert Manning.

in Normandy, a year after the battle of Waterloo, in which his father had taken part. The first Arterial Drainage Act of Ireland, passed in 1842, stipulated that drainage works should be planned and executed by the employees of the Central Government under the jurisdiction of the Board of Works. The first Commissioner for Drainage was Thomas James Mulvaney. Manning joined the department in January, 1846, as a clerk, and later became Chief Engineer for the Office of Public Works. He was responsible for the planning, design, and construction of various drainage, navigation, and harbour projects.

Manning's papers[21,22] clearly indicate that he considered himself to be a hydrologist. His major contribution to hydrology was the paper *On the flow of water in open channels and pipes*[22] which he presented to the Institution of Civil Engineers of Ireland on 4 December, 1889. Near the beginning of his paper, he gave hydrology a well deserved boost:

'Among the numerous subjects a knowledge of which is essential to the practice of the profession of the Civil Egineer, there are none more important than those which range themselves under the comprehensive title of *Hydrology*.'[23]

Like Du Buat and Bazin, he also pointed out the imperfection of the existing state of knowledge in hydrology and hydraulics:

'Even at the present day great differences of opinion exist among writers on the subject, each investigator claiming some excellence over those who preceded him, or roundly stating that the rule proposed by him is the only correct one.'[24]

In 1867, 22 years before Manning's paper was presented, Philippe Gaspard Gauckler (1826–1905), an engineer of the Ponts et Chaussées,[24] proposed[25] the following two general formulae for all types of channels, the choice of which depended on their slope:[25,26]

$$V = \lambda_1 R^{4/3} S \qquad \text{for } S < 0.0007$$
$$V = \lambda_2 R^{2/3} S^{\frac{1}{2}} \qquad \text{for } S > 0.0007.$$

In 1881, Hagen derived[27] a formula very similar to the second one of Gauckler, probably quite independently, but in doing so, he placed no limitation on the slope.[28] He had derived it from Kutter's data.

Manning was also probably not aware of the Gauckler formulae when he analysed various available experimental data and announced his belief that the expression $V = CS^{\frac{1}{2}}R^{2/3}$ may be sufficiently accurate. Obviously, this was exactly the same as the second Gauckler formula. Manning used it to determine velocities from 170 experiments comprising the observations of Bazin, Kutter, Revy, Fteley and Stearns, and Humphreys and Abbot, and found that only 25 of them differed from the actual velocities observed by more than 7%. In this regard Manning said:

'Although the formula was independently found by the author in 1885, it is proper to say that Major Allan Cunningham, R.E., states in his paper, 'Recent Hydraulic Experiments' (Proceedings of Inst. Civil Engineers, 1882), that the experimental results of Kutter's work had been recently applied by Dr. Hagen ... [who] ... deduced by the method of least squares $V = CR^{2/3}S^{\frac{1}{2}}$ 'but the probable errors computed therewith appear enormous', ...'[29]

Manning, however, was not too keen about the formula primarily because it was incorrect dimensionally, and fractional powers like 2/3 are too cumbersome for practical use.
Having thus discarded the present-day so-called Manning formula he ventured to propose a new one:

$$V = C\sqrt{Sg}\left[R^{\frac{1}{2}} + \frac{0.22}{m^{\frac{1}{2}}}(R - 0.15)\right]$$

where C = a coefficient which depends on the nature of the bed (it is not the Chézy C), and m = atmospheric pressure in terms of mercury. Assuming m to be equal to 30 inches of mercury, he reduced the formula to:

$$V = 62\ S^{\frac{1}{2}}(R^{\frac{1}{2}} + \frac{R}{7} - 0.05)\ \text{in ft/sec}$$

$$V = 34\ S^{\frac{1}{2}}(R^{\frac{1}{2}} + \frac{R}{4} - 0.07)\ \text{in m/sec.}$$

The surprising aspect of the proposed formula was the inclusion of the barometric pressure. Manning was very conscious, and rightly so, of the problem of the dimensional homogeneity of any expression describing a physical process, and only by including the term m he could satisfy the necessary condition. Like Du Buat, he concluded that some function which is 'very small, nearly constant, and a square root', is generally neglected. Moreover, only by including barometric pressure, could Manning obtain correct discharge values for water passing through pipes of small diameters. (It should be remembered that Manning was seeking a generalized formula for determining the velocity of flow in both pipes and open channels.) In 1890, Flamant saw an advance copy of Manning's paper, and recommended[30] the use of the simpler formula, $V = CR^{2/3}S^{\frac{1}{2}}$, in his book *Mechanique appliquée – hydraulique*, published in 1891. Willcocks and Holt[31] referred to the equation as the Manning formula in 1899, and that practice was followed by Buckley[32] in 1911. Thus, gradually, the rejected expression became the well-known Manning formula. Manning spent four years in perfecting his universal formula, but, ironically, the formula with which his name is presently associated, is the one that he had previously discarded as not being sufficiently practicable.

RIVER FLOW RECORDS

Even though records of high flood levels of the Nile can be traced back[33] to the dawn of civilization, river flow records prior to the seventeenth century, were, in general, qualitative rather than quantitative. For example, the available brief descriptions of the floods of the Rhone, the Loire, and the Seine rivers for the years 563, 572, and 583 A.D., respectively,[34] are primarily about the damage to properties and loss of lives and livestock.

The greatest milestone relating to river flow measurements was undoubtedly Castelli's establishment of the concept that $Q = AV$. With it flows might have been estimated even then with reasonable accuracy, but progress was rather slow. The interest in obtaining regular stage records and analysing them was aroused early in the eighteenth century. For example, the record of stage readings of the river Elbe near Magdeburg from 1727 to 1869, a period of 143 years, was published[35] and analysed by Maass, the Royal Prussian Inspector of Hydraulic Works in 1870.

In 1837, the distinguished German hydrographer, Heinrich Berg-haus,[36] published his analysis of the highest, lowest, and the mean gauge readings of the following records:

(1) the river Rhine at Emmerich (Dutch frontier), from 1770 to 1835 (66 years);

(2) the river Rhine at Cologne, from 1782 to 1835 (54 years);

(3) the river Elbe at Magdeburg, from 1728 to 1835 (107 years); and

(4) the river Oder at Küstrin, from 1778 to 1835 (58 years).

Later, Wex[37] carried out extensive analyses of river stages of five principal Central European rivers, and 'furnished unassailable proofs' that the discharges of the rivers had continually decreased over a long period of years. He, however, had a comforting thought:

'There need probably be no apprehension that the low water-surface of the Danube, Rhine, Elbe, and Vistula will ever go down to their beds, that is, that they will become partially dry, because the first two are partially fed by the ice and snow of the Alps; because the causes which create this decrease will probably not act beyond a certain point, and because the many tributary creeks and rivers which empty into these streams generally have their highest and lowest stages at different periods of time.'[37]

It is difficult to understand how the last factor came into the picture because variations of flow in all tributaries will always take place. Later, efforts to analyse the data, by applying systematic corrections to the gauge heights, invalidated Wex's theory of progressive reduction of discharge.

Jarvis presented the minimum and the maximum flow levels of the Nile as recorded by the Roda nilometer in graphical form for the period 622 to 1926 A.D. The data is reasonably complete – except for parts of the sixteenth and seventeenth centuries. It indicated an average sedimentation rate of about 10 to 15 centimeters per century.[34]

Kolupaila[38] attempted to reduce the stage records of the Memel river at Schmalleningken from the observations made by the Lithuanian Hydrometric Bureau from 1812 to 1930.[39] Details of stage measurements of the eighteenth and the nineteenth centuries have been listed by Kolupaila[40] and by the Miami Conservancy District.[41] Systematic computation of discharge was started around

the beginning of the nineteenth century. Notable among them are
the observations of Hans Conrad Escher von der Linth (1767–1823)
for the Upper Rhine, near Basel, from 1809–1821; Antoine Joseph
Chrétien Defontaine (1785–1856) for the Rhine and its tributaries
from 1820 to 1833; Giuseppe Venturoli (1768–1846) for the Tiber
at Rome from 1825 to 1836; and André Gustave Adolphe Baum-
garten (1805–1856) for the Garonne from 1837 to 1856. Probably
the discharges were computed by the various slope-velocity formulae
which abounded during the period. The first international discharge measurement was organized in November, 1867, on the
Rhine at Basle.[42]
Baron Cornelis Rudolf Theodorus Kraijenhoff (1758–1840) published[43] a comprehensive set of hydrographic and topographic
tables for Holland in 1813. This early work is of considerable value
as it gave detailed records of discharge using the slopes of the water
surfaces, gauge heights, and velocities. Velocities were determined
by finding the time required for a vertical float-pole, extending
from above the water surface to nearly the bottom of the rivers,
to travel from one base line to another. All gauge heights were
referred to a common datum.

DEVELOPMFNT OF THE RATIONAL FORMULA

Probably the first logical attempt to estimate flood flow was made
by a group of Irish engineers[44] during the period 1842 to 1847.
The method, in brief, was to design drainage channels capable of
carrying off a certain percentage of recorded maximum daily rainfall. It was assumed that the total rainfall was disposed of in three
ways: evaporation, infiltration and stream flow, with the first
two losses being constant throughout the year. Thus, it was reasoned
that if a certain percentage of total annual precipitation found its
way to the streams, a similar proportion of daily rainfall would do
likewise.

According to Dooge,[44] Samuel Roberts, in his report of December,
1843, on the river Dee, anticipated a maximum daily rainfall
value of 1.6 in., and a run-off factor of 0.4 for his design. Later, the
run-off factors were varied to take into account the various characteristics of the catchment area, particularly slope. Thus, William

Fraser, in his report on the Longford district (river Camlin), published in February 1844, considered two run-off factors for the catchment – 0.4 for the southern part having a fall of about 5 ft per mile, and 0.6 for the northern part with a slope of 10 to 20 ft per mile. It was believed that the rate of run-off from steeper catchments was higher, since the time available for percolation and evaporation losses was less. Generally, factors used for design were 0.4 and 0.6 (or $\frac{1}{3}$ and $\frac{2}{3}$) for flat and steep lands, respectively. In the beginning, the Irish engineers assumed a maximum daily rainfall value of 1.5 or 1.6 in., but later, those amounts were gradually increased. Thus, by 1847, a flood formula could be written as follows:

$$Q = 2.52 \ C.I.A.$$

where Q = design discharge in cu. ft/min, C = run-off factor, I = maximum daily rainfall (1.5 to 2 in.), and A = catchment area in acres.

The originator of the present so-called rational method was Thomas James Mulvaney (1822–1892), younger brother of the Commissioner of Drainage, William T. Mulvaney. In a paper entitled *On the use of self-registering rain and flood gauges in making observations of the relations of rainfall and of flood discharges in a given catchment*, presented to the Institution of Civil Engineers of Ireland,[45] in February, 1851, he laid the foundation of the method. Mulvaney pointed out therein the necessity of a general and uniform method of collecting precipitation data, so that it could be analysed successfully to establish practical rules. He briefly described the existing concept, and stated that the empirical approach gave results 'tolerably near the truth' within certain limits only for an average catchment that was 'neither mountainy nor very flat'.[46]

Mulvaney's statement on the procedure of estimating maximum flood discharge is valid to a great extent even today:

'After having ascertained all these facts and feeling satisfied that each of them must have an important effect on the result as to the maximum flood which he is called upon to provide for, he has no faithful guide, that I am aware of, to help him to a conclusion as to the *amount* of effect on the discharge due to each or all of these conditions; he is, in fact, left to *guess* at the result after all, and unless he happens to have had some previous experience of similar cases, his guess will probably be very wide of the truth.'[47]

For maximum discharge to occur,

'a combination of circumstances as to the fall of rain and the peculiar character of the catchment may be required, that may not occur more than once perhaps in two or three years, but which it is nevertheless necessary that he should provide for.'[48]

Mulvaney also pointed out that a long period of data on the depth and intensity of precipitation and stream flow would be necessary to determine their interrelationship.

Mulvaney can be credited with the first correct understanding of the concept of the time of concentration as applied to the rational method:

'The first matter of importance to be ascertained in the case of a small or mountainy catchment, is *the time* which a flood requires to attain to its maximum height, during the continuance of a *uniform rate of fall* of rain. This may be assumed to be the time necessary for the rain which falls on the most remote portion of the catchment, to travel to the outlet, for it appears to me that the discharge must be greatest when the supply from every portion of the catchment arrives simultaneously at the point of discharge, supposing, as above promised, the *rate* of supply to continue constant, and this length of time being ascertained, we may assume that the discharge will be the *greatest possible*, under the circumstance of a fall of rain occurring, of the *maximum* uniform rate of fall for that time. . . This question of time, as regards any catchment, must depend chiefly on the extent, form, and rate of inclination of its surface; and, therefore, one great object for investigation is the relation between these causes and their effect; so that, having ascertained the extent, form and average inclination of any catchment, we may be able to determine, in the first place, the *duration of constant rain* required to produce a maximum discharge, and consequently to fix upon the *maximum rate* of rain-fall applicable to the case.'[49]

The coefficient C could be estimated by studying the retention capacity of the soil which would primarily be dependent on the geological formation and the degree of cultivation of the catchment. Mulvaney also suggested to study a series of specially selected small and suitable catchments to determine the effect of various factors on flood run-off. Thus, the rational formula was clearly implicit in Mulvaney's paper.

In another part of the paper, the Irish engineer described an automatic rain gauge (figure 6) which would cost the princely sum of £2: £1 for the rain gauge and another £1 or so for a clock!

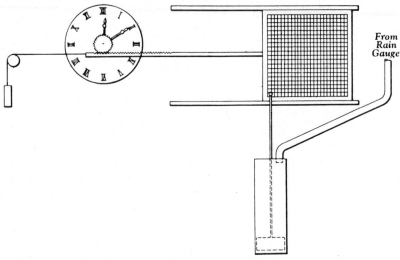

Figure 6. Automatic rain gauge of Mulvaney.

Manning, in his paper of May, 1851, on arterial drainage, pointed out that drainage works could reduce the time for floods to reach their peak, and thus increase the value of the peak flow. The maximum flow would depend on the rainfall, and the extent and circumstances of the catchment basin. He neglected the effect of evaporation and infiltration losses because

'It is evident that there may be such a state of the atmosphere that no appreciable amount of evaporation may occur during the continuance of a flood, and no matter what may be the water storing properties of the geological formation of the district they may have been completely exhausted by previous rains; and, therefore, the maximum discharge will mainly depend upon the quantity of rain, and the extent and comparative elevation of the catchment to the discharging channel.'[50]

He listed maximum daily rainfall values at five locations based on 4 to 10 years of record. Those values, however, were all exceeded during the same year as well as during the subsequent year (1852). The rational formula was later recommended by Emil Kuichling[51] in 1889, George Chamier[52] in 1898, and D. E. Lloyd-Davies[53] in 1906. It is currently known as the Kuichling formula in the United States and the Lloyd-Davies formula in the United Kingdom –

after the names of the engineers who first introduced it in their respective countries. This is a rather unfortunate circumstance, as in all fairness, it should have been named the Mulvaney formula internationally.

HERSCHEL AND THE VENTURI METER

Clemens Herschel (1842–1930) was probably born in Austria (both Boston[54] and Vienna have been mentioned as his birth place), but he was educated at Harvard, Paris, and Karlsruhe. Through his association with James Francis[55] at Lowell, Massachusetts, he acquired a keen interest in hydraulic engineering. His two main contributions to hydrology were his invention of the Venturi water meter and his work on the history of hydrology. It may be recalled that it was Herschel who took photographs of the manuscript of Frontinus at the Montecassino monastery and who located the original Chézy work – both of which he translated into English. He received the Elliott Cresson medal of the Franklin Institute for his 1898 paper on the Venturi meter, and was elected the president of both the American and Boston Societies of Civil Engineers. While at Holyoke, Massachusetts, Herschel found himself faced with the problem of finding an economical method of estimating discharge so that the local power companies could be charged an amount that was commensurate with the quantity of water they were furnished. The most obvious solution at that time was to construct a weir, but that was an uneconomical type of installation. He stated:

'There was another draft of water out of the canals, unseen by human eyes, which sorely troubled me. This was the large quantity used by the manufacturing corporations, including some 25 large paper mills, as wash-water; roughly estimated at 10 percent of the quantity used for power. This water was drawn through cast-iron pipes, most of them 20 to 24 inches in diameter, painted black on the outside, and they lay there, usually in the basement of the mill, silent as the grave, and most provokingly secretive of what was passing within their interior. Many a time did I stand beside such a pipe and exert myself to invent how to force these pipes to reveal the secret of their hidden action.
These endeavours resulted in a determination at the first opportunity to try how an apparatus like this would work: place an orifice at some point in the pipe, circular and in the form of an adjutage, from choice, and then place an

expanding cone downstream from the orifice, in order that the loss of head occasioned by the first orifice may be regained, and no material loss of head be occasioned by the whole apparatus.'[56]

Herschel noted that his water meter had been developed from a study of the combination of Bourdon's anemometer, Venturi's experiments and Boyden's turbine diffuser. His reason for naming the new water meter after the Italian, Venturi, was the

'supposition merely that as Venturi had discovered there was a sucking action at the throat, the intensity of this action would be found to have a valuable relation to the throat velocity.'[56]

Soon after Venturi meters came into general use for measuring the flow in pipes (it had its fair share of criticisms), the concept was extended (through the construction of Venturi flumes) to determine the discharge of canals and small water courses.

GROUND WATER HYDROLOGY

William Smith

In the field of ground water, one of the early applications of the principles of geology to the solution of hydrologic problems was made by the Englishman, William Smith (figure 7). Born at Churchill in Oxfordshire, in 1769, Smith is often described (somewhat unfairly to his predecessors) as the father of English geology. He described himself variously as either a geologist, a mineralogist or a civil engineer, and spent most of his fortune publishing his geological maps. In 1815, Smith bought property near Bath, England, containing freestone, but ironically, he was hopelessly wrong on both the quantity as well as the quality of the stones available. He lost heavily on the deal, and was forced to sell his excellent geological collection to the British Museum. He died at Northampton, in 1839.

Smith's main contribution to hydrology is in the field of ground water. In a paper entitled *On retaining water in the rocks for summer use*, presented in 1827 to the Yorkshire Philosophical Society,[57] he discussed the utilization of ground water for the town of Scar-

Figure 7. William Smith (by courtesy of the Geological Society of London).

borough. The paper first discussed the necessity and the desirability of conserving water; then it described a method for supplementing the town's summer water supply. A bore hole dug several years previously, to drain the land, had been found to discharge a small quantity of water. An open channel was subsequently cut there to a depth of 9 to 10 ft, and it increased the flow to about 24 hogshead per hour. It encouraged Smith to deepen the channel by another 4 ft whereupon the discharge became further increased to 50 or 60 hogshead per hour. He suspected that the water came from a confined aquifer, and suggested the 'propriety of damming up the produce of this spring' for summer and winter use, as the supply available during spring was more than adequate for that season's use. The water-bearing stratum was found to be a yellowish fine grained crumbly sand-stone occurring in thick beds with open joints. A basin, 6 ft in diameter and 4 ft deep, was excavated about 6 ft from the nearly upright edge of the rock to receive the flow of water, and pipes were laid at the bottom of the basin to carry the water to the city reservoir. In order that the water of the spring could be utilized partly or wholly as desired, Smith used four vertical pipes, each 12 in. long and removable as necessary, at the end of the pipe line. In winter, when the top of the vertical pipe was closed, the geologist was pleasantly surprised to find that the water table height at the spring rose to 14 ft, some 10 ft higher than anticipated.[58]

Darcy and Dupuit
The foundation of the theoretical aspect of ground water hydrology was laid by Darcy in his Report[11] of 1856 on the water supply system for Dijon. In one of the appendices[59] of the Report, discussing the technique of purification of water by filtration through sand, he suggested the following well-known expression which at present bears his name:

$$Q = \frac{K.A}{L} (H + L)$$

where L and A are the length and the cross-sectional area of the sample; K is a constant, and H is the head of water above the sample. The velocity V was equal to Q/A, but he did not introduce any special velocity concept or the idea of porosity.

Darcy emphasized the empirical nature of the relationship proposed which was based on careful field and laboratory observations. His main interest was to investigate the possibility of increasing the yields of wells. With the exception of the filtration tests, he did not

conduct any additional studies in the field of ground water hydrology. He completely rejected the hypothesis that rain-water was
unable to penetrate more than a few feet into the soil, and offered
a rational explanation for the seasonal variation of the productivity
of the wells. In case of artesian wells, Darcy considered the aquifer
to be analogous to a large pipe connecting two reservoirs at different
levels. Artesian wells were sort of pipes, withdrawing water from
a main line that was under pressure.

Darcy's work on ground water was extended by another Frenchman, Arsène Jules Emile Juvenal Dupuit (1804–1866), whose name
is at present synonymous with the equation for axially symmetric
flow toward a well in a pervious medium. The problem was treated
in his 1863-treatise in a chapter on seepage.[60] He was familiar with
Darcy's work on filtration, and attempted to solve the problem by
using De Coulomb's resistance law (as modified by De Prony)
for expressing open channel flow. He assumed that a mass of sand
is analogous to a collection of tiny channels to which De Prony's
equation could be applied. He further assumed that all channels
were subjected to identical conditions, and hence, the gradient and
the velocity for all 'microchannels' in a vertical section would be
the same. Since the velocity of flow through a pervious medium is
slow, he neglected the term containing v^2. Thus, he reduced the
De Prony equation for seepage to:

$$i = \eta v$$

where η is a constant depending on the nature of the soil.[61] He
pointed out the similarity between his quasi-theoretical expression
and Darcy's empirical formula.

Dupuit then deduced the following theoretical expression for the
rate of flow into a gravity well by considering an arbitrary cylindrical surface surrounding it:

$$q = \frac{\pi k (H^2 - h_0{}^2)}{\log R/r_0}$$

where, q = discharge per unit time, k = coefficient of permeability,
H = height of water table above the impervious stratum beyond
the zone of influence, h_0 = depth of water in well, R = radius of
the zone of influence, and, r_0 = radius of the well.

He also deduced two other similar equations for recharge and for artesian wells. The two basic assumptions for all three of his formulae were:

(1) the same gradient obtains at all points in a section, and,
(2) the gradient of the phreatic surface at any point is equal to the slope of the surface at that point.

The above two assumptions obviously impose serious limitations to the Dupuit formulae. But, it was rather strange that Dupuit, after having deduced the simplified formulae, persisted in disregarding his own fundamental assumptions. Unlike Darcy, Dupuit failed to conduct extensive field and laboratory investigations to check his theoretical expressions, nor did he explain any limiting value for R in his equations. If the value of R, in case of a pervious stratum were assumed to be infinity, the practicability of the formulae would have been seriously hampered. Later, Adolph Thiem suggested a reasonable value of R based on field investigations which he performed.

Dupuit's work, in spite of the limitations mentioned, greatly advanced the knowledge of ground water hydrology, even though the equation for the gravity well contained incorrect assumptions for the phreatic line and for the distribution of the piezometric head along the impervious base. It is, however, still used to calculate the discharge and/or the coefficient of permeability because the actual discrepancies between the true values and the values obtained thereby, are negligible.

Adolph Thiem

The pioneering work of Darcy and Dupuit in France, in the field of ground water hydrology, was later taken up by the Germans and Austrians.[62] The most notable German trail-blazer in this field was Adolph Thiem (1836–1908) – a civil engineer for the city of Dresden. In a paper[63] published in 1870, he made theoretical analyses of problems concerning the flow of water toward gravity wells, artesian wells and filter galleries. By adopting the necessary assumptions, he derived the same expressions as Dupuit had derived for gravity and artesian wells. He too, was as casual as Dupuit about the limitations of such assumptions. Probably both of them preferred to consider the two conditions as logical inferences, rather than as simplifying assumptions.[64]

In the same paper, Thiem considered the problem of partially penetrating gravity wells. He was, however, guilty of oversimplification, and concluded that the effect of partial penetration on yield would be inconsiderable. He also attempted to analyse the problem of non-steady seepage, but the results, he himself admitted, were of little practical use.

In later papers,[65-67] Thiem presented extensive field observations in support of his formulae. He suggested that the radius of the zone of influence in a pervious medium need not be taken as infinity. It could be restricted to the point where the drawdown is so small that it could be neglected with very little sacrifice in accuracy. Initially, Thiem attempted to measure velocity of ground water flow by injecting dye at one point and then noting the time of its subsequent appearance in an observation well. The method, however, proved to be rather inaccurate because of the tendency of the dye to disperse even in still water. Later, he resorted to the practice of injecting salt solution.[68,69] Using a salt solution of known concentration in still water, he determined the concentration at various points after some time. Armed with the calibration chart, he was able to apply the necessary corrections by determining the salt content in an observation well, at a fixed distance from the injection point and after the elapse of a specific period of time. Thiem wrote extensively in the field of ground water hydrology. Probably his greatest contributions were the stress he placed on experimental techniques, and the efforts he made to reconcile theoretical and field observations.

Forchheimer and Slichter

One of the greatest contributors to the field of ground water theories, during the late nineteenth century and the earlier half of the twentieth century, was undoubtedly Philip Forchheimer (1852–1933). A native of Vienna, and a professor of hydraulics at Aachen and later at Graz, Forchheimer, for the first time, applied advanced mathematics to this subject. One of his major contributions was a determination of the relationship between equipotential surfaces and stream lines. The analytical method on which flow net principle is based was discussed in the first edition of his book on hydraulics[70] published in 1914. This was certainly not the first work published

on the subject, as Richardson[71] had already published a paper in
1908, in England, quite independently of Forchheimer, but Forchheimer's very earliest paper on ground water[72] published in 1886
makes it clear that the idea had begun to form in his mind at that
time.

Holzmüller, in 1882, had applied the technique of conformal
mapping to heat flow problems,[73] and this inspired Forchheimer
to approach ground water flow analyses in a similar fashion.
Starting with Darcy's law and Dupuit's assumptions, he arrived[72]
at Laplace's equation for the phreatic surface for gravity flow in a
pervious stratum underlain by a horizontal impervious base:

$$\frac{\delta^2 z}{\delta x^2} + \frac{\delta^2 z}{\delta y^2} = 0$$

where x and y are the co-ordinates of the point in question on a
horizontal plane, and z is the elevation of the phreatic surface above
a horizontal impervious base.

He pointed out that the above equation is valid for the movement
of ground water at great depths, but for shallow depths, where
changes from point to point were appreciable, it was modified to:

$$\frac{\delta^2 (z^2)}{\delta x^2} - \frac{\delta^2 (z^2)}{\delta y^2} = 0.$$

Forchheimer was not only the first man to indicate the applicability
of Laplace's equation to the phenomena of ground water flow, but
he was also the first to offer a clear explanation of the Dupuit
assumptions.[74]

Forchheimer introduced the theory of functions of a complex
variable to analyse the gravity flow toward a group of wells. He
used the concept of an imaginary equivalent single well which had
the same rate of discharge as the group of wells. The proposed
equation for a steady state flow assumed that the yield would be
the same from each well, and it did not impose any restriction on
the equivalent single well, except perhaps of an implied requirement
of a roughly central location.

Forchheimer was also the first to apply the method of mirror images
to the ground water flow problems.[74] He analysed the case of a
gravity well near a river, both being underlain by a continuous

impervious base. The ground water table at the well was assumed to be at the same level as the water surface of the river. For his analysis, Forchheimer replaced the river by a continuous pervious stratum with an imaginary recharge well located as the mirror image of the given well with reference to the river bank. The imaginary well was capable of supplying as much water to the surrounding pervious medium as the real well was capable of removing. He then introduced the feature of natural ground water flow to the river, and using the method of mirror images again, determined the critical distance between a river and a well, beyond which no water would be contributed by the river to the well.

In the above cases, Forchheimer assumed that the wells penetrated to the underlying impervious base, but he also considered partially penetrating gravity wells. The approach was semiempirical, and the resulting equation was rather formidable.[74]

Forchheimer had an excellent mathematical background. He successfully used the methods of conformal mapping, mirror images, complex variables, and potential theory to solve problems in ground water hydrology, and, in nearly all cases, he was the first man to have used them. With regard to Forchheimer's contribution to hydrology, Terzaghi had this to say:

'In my opinion his contribution accomplished more in the line of clarifying our ideas concerning the movement of ground water than those of all the other contemporaneous hydrologists of Europe combined.'[75]

Much of Forchheimer's work was duplicated in the United States by Charles Sumner Slichter (1864–1946). Slichter's knowledge of European developments in the field of ground water hydrology was very limited, and apparently he was not even aware of Forchheimer's existence. Slichter derived an analytical solution for discharge from artesian wells which proved to be identical to the Dupuit–Thiem solution. Slichter, like Forchheimer, was impressed by Holzmüller's work, and successfully applied Laplace's equation to conformal transformation, and to the potential theory to ground water problems.[76]

Slichter suggested an elaborate expression for flow of water through a vertical column of soil:

$$Q = 1.0094 \; \frac{\Delta p \cdot d^2 \cdot A}{\mu h k}$$

where Q = rate of flow in cc/sec, p = difference in 'pressure' at the two ends of the column in cm of water, d = mean diameter of soil grains in cm, A = cross-sectional area in cm², μ = coefficient of viscosity of water in g.sec/cm², and k = a constant depending on the porosity and geometrical characteristics of the medium.

He realized that the expression was a form of Darcy's law but, unlike Darcy, who used a single constant k, he divided it into its various constituents. The form of the equation makes it obvious that Slichter had used the Hagen–Poiseuille equation for its derivation. Slichter used a circuit containing electrodes in observation wells and an ammeter to obtain ground water flow velocities. By injecting an electrolyte into the ground water, he could detect the rate of seepage between two points from the ammeter readings (which were dependent on the electrolyte concentration in the water). Even though a major part of Slichter's work had already been done in Europe, his contribution to the development of the science of ground water hydrology should be recognized. He did his work independently, and was largely instrumental for the advancement of the subject in America. He undoubtedly made Americans more ground water conscious.

BEARDMORE AND MANUAL OF HYDROLOGY

The book *Manual of hydrology*[77] by Nathaniel Beardmore, published in 1862, was the first work in the English language that began to approach the subject of hydrology as it is known today. It was an enlarged and revised version of his earlier work, entitled *Hydraulic tables*, published in 1850. Beardmore, born in 1816 at Nottingham, England, was responsible for the planning and design of various railways, harbours, bridges, drainage, and waterworks. He was elected the president of the Royal Meteorological Society, in 1861, and was an advisor concerning water supply schemes for cities as far apart as Edinburgh and Glasgow to Moscow and Odessa.[78] He died in 1872.

Beardmore did not suggest any new hydrologic principles. His main contribution to the subject was to popularize it. This he accomplished through his book – an excellent compilation of the state of knowledge thereon in tabular form.[79] He commented that:

> 'refined but practical questions of surface slope and velocity of water and, above all, of the volume accompanying a given fall and velocity or certain known rainfall, were subjects almost untouched (until publication of the first edition of this volume); the source or supply of water in reference to the amount of rain was a subject which only a few canal and waterworks engineers had investigated; and they were not much disposed in olden times to communicate the practical experience acquired by the hard labour of years.
> Hydrological science embraces the widest conditions; not only has climate to be considered, but the elevation, inclination, and geological formation of the substratum. Practical construction requires great previous experience, when the science has to be applied; for instance in drainage and waterworks. . .'[77]

The book, intended to be a practical manual for everyday use, was divided into four parts: hydraulic and other tables, rivers and flow, tides and rainfall. It contained hydrologic data from various parts of the world primarily for the use of the British consulting engineers who, towards the latter half of the nineteenth century, were involved in planning and design of water resource projects on an international scale.

OTHER DEVELOPMENTS

Toward the latter half of the nineteenth century, the United States Geological Survey and its antecedent organizations began a systematic collection and publication of the average daily discharge values for representative streams throughout the entire nation. Charles Ellet, Jr. (1810–1862), a hero of the Civil War, was a real pioneer of hydrometry.[80,81] He was probably the first to tabulate records of the daily discharges based on actual velocity measurements at various stages for the Ohio river near Wheeling, West Virginia. The Geological Survey was created in 1879, and, by 1906, according to Nathan C. Grover (1868–1956), Chief Hydraulic Engineer of the Survey during the period 1903 to 1939:

> 'stream gaging was nation wide; investigations of underground waters were being successfully made in both the East and West; Gilbert had begun his monumental

work on the transporting capacity of flowing water; Slichter had pioneered in measuring the rate of motion of water through the ground. . . Congressional authority had been obtained for the preparation of reports on the best method of utilizing the water resources. . . the complete yearly records of the gaging station had been brought into one publication; studies had been made of the essential accuracy of streamflow records; Murphy had investigated the reliability of the current meter; progress was being made in obtaining winter records. . . This was a record of accomplishment that would be hard to equal in such a relatively short period.'[82]

Also towards the later half of the nineteenth century, various simple flood formulae of the type
$$Q = C \cdot A^n$$
were proposed, where C was a coefficient, n was an index (both depending on locality), and A was the drainage area in acres. Probably the earliest such formula was proposed by Colonel C. H. Dickens[83] based on his observation in India:
$$Q = C \cdot A^{0.75}$$
where C varied from 1.56 to 17.2.
Chow[84] has excellently summarized various flood formulae proposed during the period of 1860 to 1950.
Rippl, in 1883, suggested a method[85] for determining the minimum effective storage required so that no water shortage occurs during the time period under consideration. The method was based on residual mass diagram, and assumed that both inflows and outflows were known functions of time. It was pointed out during the discussion of the paper that the suggested method had been used by many engineers several years prior to the publication of Rippl's paper.
At the forefront among the Italian hydrologists of this period were Giuseppe Venturoli (1768–1846), Giorgio Bidone (1781–1839), and Elia Lombardini (1794–1878). Bidone was probably the first man to analyse the phenomenon of hydraulic jump on a systematic basis, and it is still known as the 'jump of Bidone' in Italy. He conducted experiments on discharge over weirs, and his results were published in the *Mémoires de l'Academie des Sciences de Turin* in 1820, 1826 and 1827. Venturoli's contribution was to derive the elementary backwater equation for rectangular channels in 1823. He was able to plot various reaches of the surface profile by graphical

integration of the differential equation. He also analysed the flow of the river Tiber over a number of years. The last, but not least, of this group was Lombardini, who in a series of articles and treatises described the hydrology of the river Po, and discussed various flood control programmes. He pointed out the possibility of the occurrence of higher floods due to the deforestation of mountain sides – thus, causing more rapid run-off. He analysed statistically the monthly flows of several Italian rivers.

The Hagen–Poiseuille equation for flow through circular tubes was derived during the nineteenth century, and to a certain extent it contributed to the development of ground water hydrology. The men who developed it from experimental data, were the German hydraulic engineer Gotthilf Heinrich Ludwig Hagen (1797–1884), and the Paris physician, Jean Louis Poiseuille (1799–1869). Strangely enough, the analytical solution was also proposed independently at about the same time by two physicists, Franz Newmann (1798–1895) of Königsberg and Eduard Hagenbach (1830–1910) of Basle. However, it was Hagenbach who named his resistance law for laminar flow after Poiseuille, and that identification, somewhat unfairly, still persists.

Two European meteorologists deserve special mention for their effort to systematize precipitation measurements. They are George James Symons (1838–1900) and Johann Georg Gustav Hellmann (1854–1939), and both of them made serious studies of the history of various aspects of meteorology (see chapter 12). Symons spent nearly forty years of his life co-ordinating the various rainfall observations throughout Britain, and it was primarily because of his efforts that the first volume of *English rainfall* was published in 1860–1861. The annual precipitation values have been published regularly ever since. In the twentieth annual volume of the *British rainfall*, 1880, he pointed out that:

'There is no point in the study of rainfall of greater interest and practical utility, than the accurate determination of the average annual fall.'[62]

Symons realized that it is all too easy to obtain an approximate measurement of rainfall, but it is progressively more difficult to improve on it, and extremely difficult to obtain and verify an absolute rather than a conventionally standard measurement. He suggested that even if such a method were found, it would probably not be practicable to adapt it to a national net.

The use of current meters became popular toward the latter half of the nineteenth century. The origin of current meters is rather obscure, but their antecedents can be traced to the designs of anemometers, windmills, water wheels, or ship's logs[86] – which were available long before the first current meter made its debut. Some of the current meters used during this period were devised by Daniel Farrand Henry[86] (1833–1907), General Theodore G. Ellis (1829–1883), William Gunn Price[87] (1853–1928) and Clemens Herschel. The number of revolutions of the meter wheel of the earlier instruments were indicated by a mechanical counting device, and hence it was necessary to raise the meter out of the water for each reading. In about 1860, Daniel Farrand Henry of the United States Lake Survey, Detroit, Michigan, invented an electrical facility for recording the number of revolutions of the wheel while the meter was still in water, thus making it less cumbersome than the earlier models.[86]

Other contributors to the development of hydrology in the nineteenth century were Jean Baptiste Belanger (1789–1874), Marie Francois Eugène Belgrand (1810–1878), Emmanuel Joseph Boudin (1820–1893), Jacques Antoine Charles Bresse (1822–1883), Alfonse Fteley (1837–1903), John Fletcher Miller, Abbé Paramelle (1790–1875), Jean-Claude Barré de Saint Venant (1797–1886), Frederic Pike Stearns (1851–1919), and Julius Weisbach (1806–1871).

CONCLUSION

The major achievement of the nineteenth century was the firm establishment of the principle of conducting experimental investigations either to establish a theory or to determine an empirical relationship. It is true that often the various formulae proposed with regard to the same phenomenon differed substantially from one another, but that took place mainly because of a tendency to generalize from a limited amount of experimentation. This criticism is especially valid with regard to many of the equations proposed for determining the flow in open channels.

The two major developments in hydrology of this period were in the field of streamflow observations and ground water. The perfection of current meters revolutionized the methods of evaluating river discharges, and the availability of automobiles greatly reduced the transportation problems generally associated with such activities. From the time of Leonardo da Vinci to about 1870, surface floats were extensively used to determine the velocity of flowing water. The mechanical current meters, that were available, were extremely cumbersome to use, and, hence, it is not surprising to find that they were not preferred until the number of revolutions could be counted through the use of an electrical device.

The United States Geological Survey deserves primary credit for the innovation of systematic collection and publication of streamflow data for representative streams across the nation. From an overall viewpoint, it has probably become the greatest boon the engineering profession has ever been handed.

For the first time in history, attempt was also made to estimate the design flood for a catchment on a logical basis. Credit for this must go to the group of enterprising Irish civil engineers – especially Mulvaney.

In the field of ground water hydrology, the marriage between geology and hydrology was performed by William Smith, who has often been called the father of English geology. The lead, however, was soon taken over by the French engineers who became very active in this field of geohydrology. The most notable among those engineers were Darcy and Dupuit. A serious difficulty of this period was the limited exchange of information between investigators; it was responsible, for example, for Thiem having duplicated the work of Dupuit – even after a reasonable time lag. This circumstance is rather difficult to explain because the various works in the field of making discharge measurements do not seem to have been confined to any national boundaries. For instance, Kolupaila lists 39 reviews and discussions of the Humphreys and Abbot Report, out of which no less than 18 are in foreign languages. From that it would seem that during this period surface water measurements were considered to be of greater importance and of more practical use than the investigations which were being made in geohydrology.

REFERENCES

1. EYTELWEIN, J. A., Handbuch der Mechanik fester Körper und der Hydraulik. Berlin (1801).
2. EYTELWEIN, J. A., A summary of the most useful parts of hydraulics, an English extract of ref. 1 by Young. In: Tracts on hydraulics, edited by T. Tredgold. London, J. Taylor (1826).
3. EYTELWEIN, J. A., Handbuch der Hydrostatick. Berlin (1826).
4. DE COULOMB, C. A., Expériences destinées à déterminer la cohérence des fluides et les lois de leur résistance dans les mouvements très lents. Mémoires de l'Institut National des Sciences et Arts, vol. 3 (1800).
5. GIRARD, P. S., Mémoires de l'Institut National des Sciences et Arts, vol. 9 (1803) p. 246.
6. RENNIE, G., Report on the progress and present state of our knowledge of hydraulics as a branch of engineering. Report of the British Association for the Advancement of Science, part I, 3rd meeting, Cambridge, 1833 and part II, 4th meeting, Edinburgh, 1834. London, J. Murray (1834) pp. 153–184; (1835) pp. 415–512.
7. DE PRONY, G. C. F. M. R., Recherches physico-mathématiques sur la théorie des eaux courantes. Paris (1804).
8. COURTOIS, E., Principes d'hydraulique rationale applicables aux courants naturels tels que les rivières et les fleuves. Paris, Mallet-Bachelier (1850).
9. LAHMEYER, J. W., Erfahrungs-Resultate über gleichförmige Bewegung des Wassers in Flussbetten und Kanälen. Braunschweig (1845).
10. DE ST. VENANT, J. C. B., Formules et tables nouvelles pour la solution des problèmes relatifs aux eaux courantes. Annales des Mines (1851).
11. DARCY, H. P. G., Les fontaines publiques de la ville de Dijon. Paris, V. Dalmont (1856).
12. LELIAVSKY, S., River and canal hydraulics. London, Chapman and Hall (1965) p. 6.
13. DARCY, H. P. G. and H. E. BAZIN, Recherches hydrauliques. Paris (1865).
14. BAZIN, H. E., Etude d'une nouvelle formule pour calculer le débit des canaux découverts. Paris, P. Vicq-Dunod (1898). (Extrait des Annales des Ponts et Chaussées, 1897.)
15. DAVIES, S. J. and C. M. WHITE, A review of flow in pipes and channels. Engineering *128* (1929) 69–72; 98–100; 131–132.
16. HUMPHREYS, A. A. and H. L. ABBOT, Report upon the physics and hydraulics of the Mississippi river. Philadelphia (1861).
17. KOLUPAILA, S., Bibliography of hydrometry. Notre Dame, Indiana, University of Notre Dame Press (1961) pp. 253–258.
18. Ref. 16 2nd ed. Washington, D.C., Government Printing Office (1876) p.187.
19. GANGUILLET, E. O. and W. R. KUTTER, Versuch zur Aufstellung einer neuen allgemeinen Formel für die gleichförmige Bewegung des Wassers in Kanälen und Flüssen. Bern, Druck von Lang (1877).

20. MERRIMAN, M., Treatise on hydraulics, 8th ed. New York, John Wiley & Sons (1905) pp. 281–283.

21. MANNING, R., Observations on subjects connected with arterial drainage. Proceedings of the Institution of Civil Engineers of Ireland, May (1851) 90–104.

22. MANNING, R., On the flow of water in open channels and pipes. Proceedings of the Institution of Civil Engineers of Ireland 20 (1890) 161–206.

23. Ibid., p. 161.

24. ROUSE, H. and S. INCE, History of hydraulics. New York, Dover Publications (1957) p. 179.

25. GAUCKLER, P. G., Etudes théoriques et pratiques sur l'écoulement et le mouvement des eaux. Comptes Rendus du L'Academie des Sciences 64 (1867) 821.

26. GAUCKLER, P. G., Du mouvement de l'eau dans les conduites. Annales des Ponts et Chaussées 15 (1868) 229–281.

27. HAGEN, G. W., Neuere Beobachtung über die gleichförmige Bewegung des Wassers. Zeitschrift für Bauwesen 31 (1881) 403–408.

28. POWELL, R. W., The origin of Manning formula. Journal of the Hydraulics Division, ASCE 94 (1968) 1179–1181.

29. MANNING, R., Ref. 22, p. 177.

30. FLAMANT, A., Méchanique appliquée – hydraulique. Paris (1891).

31. WILLCOCKS, W. and R. HOLT, Elementary hydraulics. Cairo, National Printing Office (1899).

32. BUCKLEY, R. B., Design of channels for irrigation and drainage. London, Spon (1911).

33. BISWAS, ASIT K., The Nile: its origin and rise. Water and Sewage Works 113 (1966) 282–292.

34. JARVIS, C. S., Flood-stage records of the river Nile. Proceedings ASCE 61 (1935) 803–812.

35. MAASS, Die Wasserstände der Elbe in den Jahren 1727 bis 1870. Zeitschrift für Bauwesen, Berlin (1870) 496–502.

36. BERGHAUS, H. C. W., Allgemeine Länder- und Völkerkunde, 2nd vol., Umrisse der Hydrographie. Stuttgart (1837).

37. WEX, G. V., First treatise on the decrease of water in springs, creeks, and rivers, translated by G. Wietzel. Washington, D.C., Government Printing Office (1881) p. 27.

38. KOLUPAILA, S., Die Bestimmung des Abflüsses des Memelströmes (Nemunas, 1812–1932). IV Hydrologische Konferenz der Baltischen Staten, Leningrad, September (1933).

39. DAVENPORT, R. W., Long records of river flow. In: Hydrology, edited by O. E. Meinzer. New York, Dover Publications, Inc. (1942) pp. 499–506.

40. KOLUPAILA, S., op. cit., pp. 12–15.

41. MIAMI CONSERVANCY DISTRICT, Storm rainfall of eastern United States. (Revised Technical Report, part 1, Dayton Ohio 1919 pp. 314–328.

42. GREBENAU, H., Die Internationale Rheinstrom-Messung bei Basel vorgenom-men am 6–12, November 1867. Munich, J. Lindauer (1873).

43. KRAIJENHOFF, C. R. T., Recueil des observations hydrographiques et topo-graphiques faites en Hollande. Amsterdam, Chez Doorman et Comp. (1813).

44. DOOGE, J. C. I., The rational method for estimating flood peaks. Engineering *184* (1957) 311–313; 374–377.

45. MULVANEY, T. J., On the use of self-registering rain and flood gauges in making observations of the relations of rainfall and flood discharges in a given catchment. Proceedings of the Institution of Civil Engineers of Ireland *4* (1850–1851) 18–31.

46. *Ibid.*, p. 20.

47. *Ibid.*, pp. 19–20.

48. *Ibid.*, pp. 20–21.

49. *Ibid.*, pp. 23–24.

50. MANNING, R., Ref. 21, p. 94.

51. KUICHLING, E., The relation between the rainfall and the discharge of sewers in populous districts. Transactions, ASCE *20* (1889) 1–56.

52. CHAMIER, G., Capacities required for culverts and flood openings. Procee-dings of the Institution of Civil Engineers *134* (1898) 313–323.

53. LLOYD-DAVIES, D. E., The elimination of storm water from sewerage systems. Proceedings of the Institution of Civil Engineers *164* (1906) 41–67.

54. RICHARDSON, C. G., The 50th year of a great invention. Water Works Engi-neering *90* (1937) 1326–1329; 1382.

55. HERSCHEL, C., A farewell word on the Venturi meter. Engineering News-Record *102* (1929) 636–637.

56. HERSCHEL, C., The Venturi water meter: an instrument making use of a new method of gauging water. Transactions, ASCE *17* (1887) 228.

57. SMITH, W., On retaining water in the rocks for summer use. Philosophical Magazine, New Series *1* (1827) 415.

58. SHEPPARD, T., William Smith: his maps and memoirs. Proceedings of the Yorkshire Geological Society *19* (1917) 189–191 (published as a book in 1920 by A. Brown & Sons Ltd., Hull).

59. DARCY, H. P. G., Ref. 11, Appendice – note D, translated by J. J. Fried. Water Resources Journal, AWRA *1* (1965) 5–11.

60. DUPUIT, A. J. E. J., Etudes théoriques et pratiques sur le mouvement des eaux courantes, 2nd ed. Paris, Carilian-Goeury (1863).

61. DUPUIT, A. J. E. J., Traité théorique et pratique de la conduite et de la dis-tribution des eaux. Paris, Carilian-Goeury (1854).

62. BISWAS, ASIT K., Development of hydrology in the nineteenth century. Water Power *21* (1969) 16–21.

63. THIEM, A., Die Ergiebigkeit artesischer Bohrlöcher, Schachtbrunnen, und Filtergallerien. Journal für Gasbeleuchtung und Wasserversorgung *14* (1870) 450–467.

64. HALL, H. P., A historical review of investigations of seepage toward wells. Journal of the Boston Society of Civil Engineers *41* (1954) 251–311.

65. THIEM, A., Beitrag zur Kenntnis der Grundwasserverhältnisse im norddeutschen Tieflande. Journal für Gasbeleuchtung und Wasserversorgung *24* (1881) 686–695.

66. THIEM, A., Der Versuchsbrunnen für die Wasserversorgung der Stadt München. Journal für Gasbeleuchtung und Wasserversorgung *23* (1880) 156–164.

67. THIEM, A., Resultate des Versuchsbrunnens für die Wasserversorgung der Stadt Strassburg. Journal für Gasbeleuchtung und Wasserversorgung *19* (1876) 707–718.

68. THIEM, A., Neue Messungsart natürlicher Grundwassergeschwindigkeiten. Journal für Gasbeleuchtung und Wasserversorgung *31* (1888) 18–28.

69. THIEM, A., Verfahren für Messung natürlicher Grundwassergeschwindigkeiten. Polytechnische Notizblatt *42* (1887).

70. FORCHHEIMER, P., Hydraulik. Leipzig, Teubner (1930).

71. RICHARDSON, L. F., A freehand graphic way of determining streamlines and equipotentials. London, Edinburgh and Dublin, Philosophical Magazine and Journal of Science *15* (1908) 237–269.

72. FORCHHEIMER, P., Über die Ergiebigkeit von Brunnen-Anlagen und Sickerschlitzen. Zeitschrift der Architekten- und Ingenieur-Verein *32* (1886) 539–564.

73. HOLZMÜLLER, G., Einführung in die Theorie der Isogonalen Verwandtschaften und der Konformen Abbildungen. Leipzig (1882).

74. FORCHHEIMER, P., Grundwasserspiegel bei Brunnenanlagen. Zeitschrift der österreichische Ingenieur- and Architekten-Verein *50* (1898) 629–635; 645–648.

75. Quoted by O. E. MEINZER in: Hydrology. New York, Dover Publications, Inc. (1942) p. 23.

76. SLICHTER, C. S., The motion of underground waters. Water-Supply Paper 67, US Geological Survey Washington, D.C., Government Printing Office (1902).

77. BEARDMORE, N., Manual of hydrology. London, Watelow & Sons (1862).

78. ANONYMOUS, Nathaniel Beardmore, Memoir. Proceedings of the Institution of Civil Engineers *32* (1872–1873) 256–264.

79. BISWAS, ASIT K., A short history of hydrology. Proceedings of the International Seminar for Hydrology Professors, University of Illinois, Urbana (1969).

80. KOLUPAILA, S., Early history of hydrometry in the United States. Journal of the Hydraulics Division, ASCE *86* (1960) 1–51.

81. FOLLANSBEE, R., A history of the water resources branch of the United States Geological Survey to June 30, 1919. Geological Survey, Washington, D.C. (1938) p. 41.

82. GROVER, N. C., Progress in branch activities to June 30, 1906. Geological Survey, Washington, D.C. (1938) p. 183.

83. DICKENS, C. H., Flood discharge of rivers. Professional Papers on Indian Engineering. Roorkee, India, Thomson College Press (present University of Roorkee) *2* (1865) 133–136.

84. CHOW, V. T., Hydrologic determination of waterway areas. Bulletin No. 462. Engineering Experiment Station, University of Illinois, Urbana (1962) pp. 70–79.

85. RIPPL, W., The capacity of storage reservoirs for water supply. Proceedings of the Institution of Civil Engineers *71* (1883) 270–278.

86. FRAZIER, A. H., Daniel Farrand Henry's cup type 'telegraphic' river current meter. Technology and Culture *5* (1964) 541–565.

87. FRAZIER, A. H., William Gunn Price and the Price current meters. Contributions from the Museum of History and Technology, Paper 70, Smithsonian Institution, Washington D.C. (1967).

Index